Making Geometry

MAKING GEOMETRY

Exploring Three-Dimensional Forms

Jon Allen

Floris Books

First published in 2012 by Floris Books

British Library CIP Data available

ISBN 978-086315-914-5

Printed in Poland

Contents

Acknowledgments **7**

Introduction **8**

A Family Tree of Regular and Semi-Regular Convex Polyhedra **12**

Circles and Spherepoints **14**

Constructing Small 3D Stick Models **16**

The Five Platonic Solids

- Before you start 16
- Tetrahedron 18
- Cube (Hexahedron) 19
- Octahedron 20
- Icosahedron 21
- Dodecahedron 22
- Further Demonstrations 24

Paperfolds for the Five Platonic Solids **28**

- Before you start 29
- Net for Tetrahedron 30
- Net for Octahedron 32
- Net for Icosahedron 34
- Net for Cube 36
- Net for Dodecahedron 38

Paperfolds for the Thirteen Archimedean Solids **42**

- Net for Truncated Tetrahedron 43
- Net for Truncated Octahedron 45
- Net for Cuboctahedron (Dymaxion) 48
- Net for Truncated Cube 52
- Net for Rhombicuboctahedron 58
- Net for Truncated Cuboctahedron 62
- Net for Truncated Icosahedron 67
- Net for Icosidodecahedron 71
- Net for Truncated Dodecahedron 74
- Net for Snub Cube 78
- Net for Truncated Icosidodecahedron 81
- Net for Rhombicosidodecahedron 87
- Net for Snub Dodecahedron 90

Paperfolds for Other Polyhedra **93**

- Platonic Compounds: 94
- Two Tetrahedra 94
- Cube + Octahedron 96
- Icosahedron + Dodecahedron 98
- Euclid's Method for Making a Dodecahedron 100
- Small Stellated Dodecahedron 102
- Great Dodecahedron 105
- Great Stellated Dodecahedron 108
- Net for Rhombic Dodecahedron 112
- Net for Rhombic Triacontahedron 116

Appendix A: Some Mathematical Terms **118**

Appendix B: Plain Nets for All Platonic and Archimedean Solids **120**

Appendix C: Data Table for the Platonic and Archimedean Solids **133**

Appendix D: Recommended Reading and Resources **134**

Index **135**

Acknowledgments

My thanks to Floris Books for asking for a second book following on from my *Drawing Geometry,* to Helena Waldron and their production team, and once again to their editor Christopher Moore.

My thanks to my wife Clare for her invaluable support and assistance throughout.

My thanks to Richard Ivey for the photographs on pages 16 and 17.

All other photographs were taken by the author.

All the drawings and diagrams in the book were hand drawn by the author.

A note on the arrangement of the solids

The order of the Platonic Solids (starting on page 30) is based on the degree of difficulty of the drawing of their nets.

The order of the Archimedean Solids (starting on page 43) follows the increasing number of their faces – from 8 to 92.

Introduction

God is a circle whose centre is everywhere and whose circumference is nowhere.

Variously attributed

Larry Merculieff, indigenous Aleut leader from the Pribilof Islands in the Bering Sea off Alaska, talks inspiringly of how the traditional hunters of the local islands were able to sit on the seashore for hours waiting for the sea lions to appear. They would sit for hours without losing attention, without 'zoning out', without going into a daydream – without sending a text or making a phonecall! – maintaining an alert relaxed attentiveness and deep attunement to their surroundings.

Arguably this experience of being 'one with everything', part of a greater wholeness, is as natural to us as the curiosity that directs our attention to the many isolated details of life. And this quality of being and attentiveness is very much part of making geometry.

The root of the geometry explored in this book is the cosmological geometry taught by Pythagoras and Plato, and by a chain of teachers including, in our own times, Keith Critchlow, with whom I worked for twenty years. As an insight into the way the world is made, ordered and patterned, geometry is not exclusive to Western Civilisation, but found a particularly rich expression here. Though much of the deeper knowledge has been lost, certain traces were seeded into our cultural heritage in a way that has preserved it for 2,500 years. An example of this is what every school boy and girl knows as Pythagoras' Theorem ($a^2 + b^2 = c^2$), easily remembered as the '3-4-5 triangle'.

Pythagorean or Platonic Geometry can be seen as a study of Wholeness. In two dimensions geometry begins with a circle, and in three dimensions with a sphere. Both circle and sphere begin with a centre – a point without dimension, but the origin or place of conception from which all points on the circumference of the circle or surface of the sphere are generated by means of a measure (the 'radius'). In two dimensions, a circle stands for the unity, singularity and completeness of wholeness; as does the sphere in three dimensions. Both are composed of an infinite series of points (all possibility) equidistant from and directly related to the single unique centre (the 'origin').

From the infinity of points on the circumference of a circle a selection of equally spaced points can be chosen which will create regular polygons. Similarly, on a sphere patterns of equally spaced points can be chosen which will create the regular and semi-regular polyhedra – 'regular' meaning that the faces are all the same regular polygon; 'semi-regular' meaning that the faces are all regular polygons, but of more than one type. These regular and semi-regular shapes – called the Platonic and Archimedean solids – are the fundamental elements of geometry, and are the subject of this book.

The knowing of the traditional hunters of the Pribilof Islands was not head-knowledge but heart-knowledge – an active participating in and experiencing of the ever-present moment. Drawing Geometry (the subject of the author's first book) and Making Geometry are activities that have more to do with developing the knowing of the heart than accumulating knowledge in the head. They are also deeply practical and grounding – about being present. I hope that the enjoyment you gain – a good indication of the heart's engagement – will be worth the patience entailed.

Jon Allen, March 2012

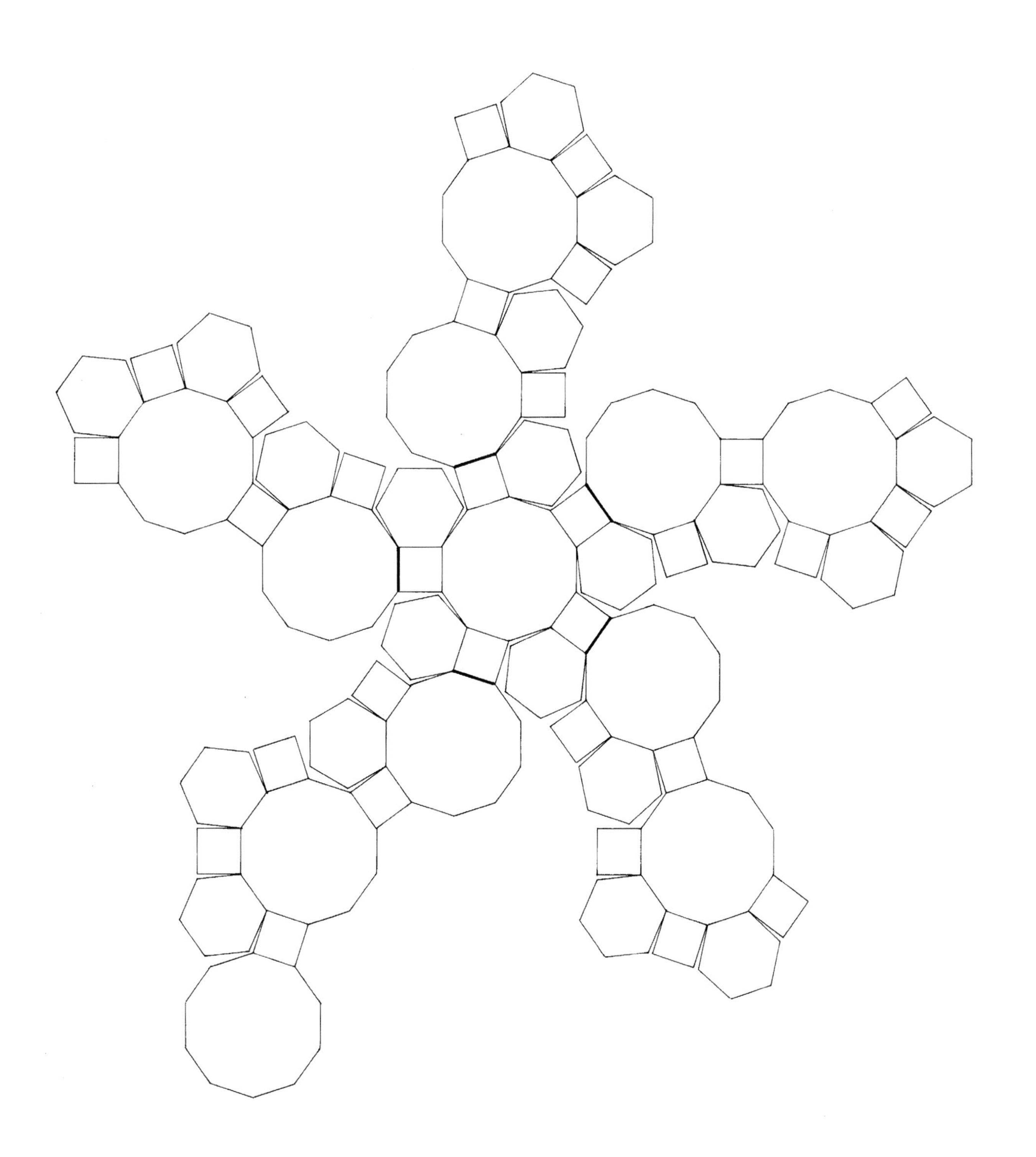

A Family Tree of Regular and Semi-Regular Convex Polyhedra

The members of this family are related by symmetry. The table is best read from left to right across the two pages.

	PLATONIC SOLIDS		ARCHIMEDEAN SOLIDS	
Octahedral 2, 3, 4-fold symmetry	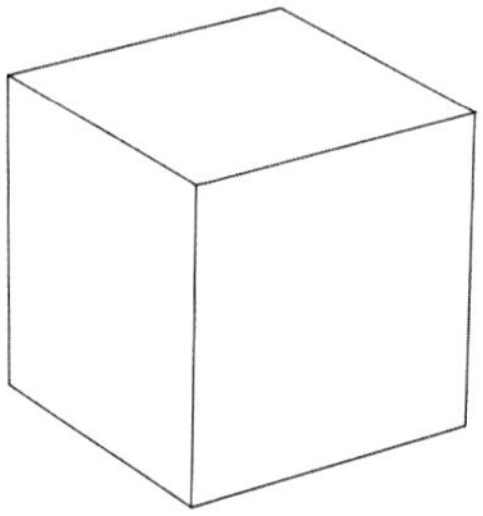 Cube	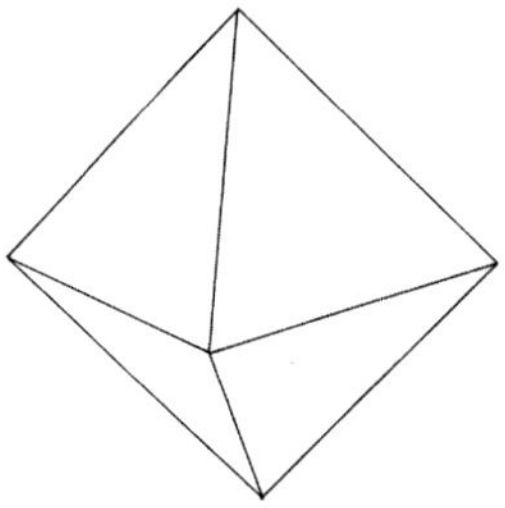 Octahedron	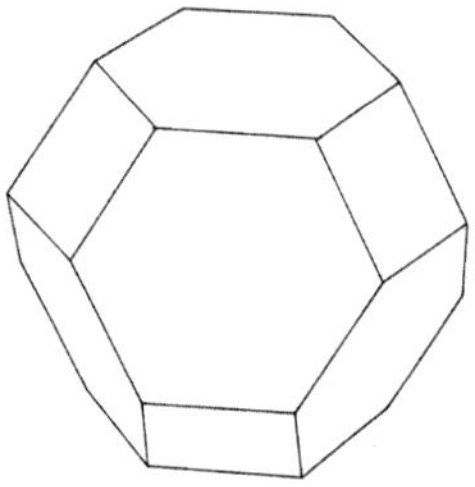 Truncated Octahedron	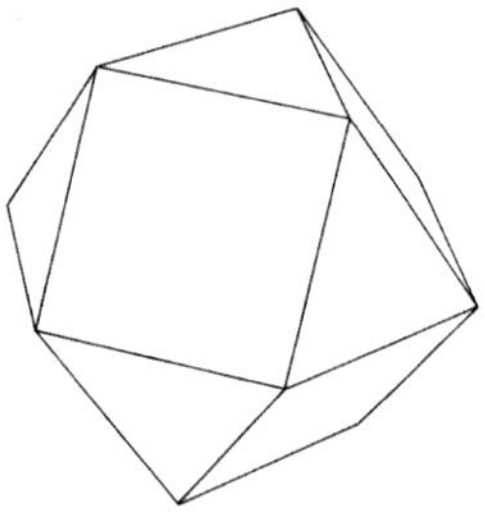 Cuboctahedron
Tetrahedral 2, 3-fold symmetry		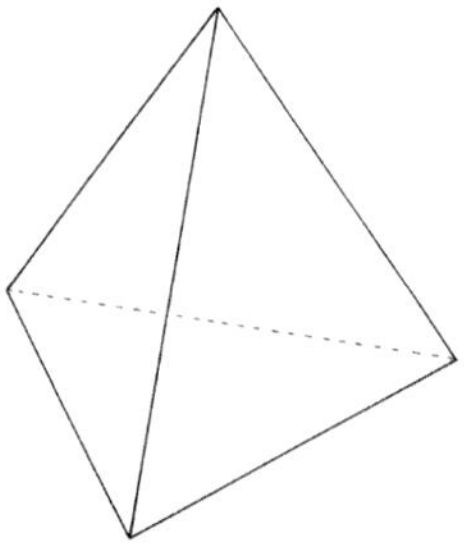 Tetrahedron	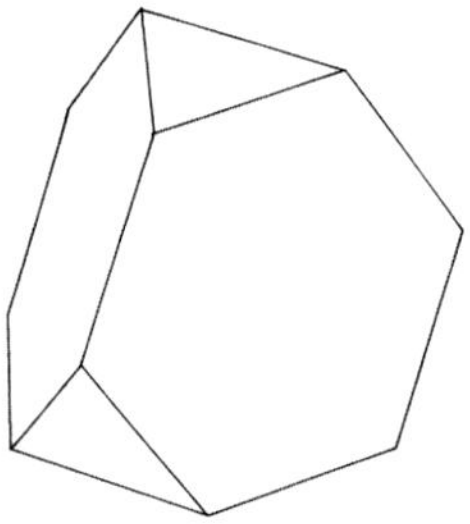 Truncated Tetrahedron	
Icosahedral 2, 3, 5-fold symmetry	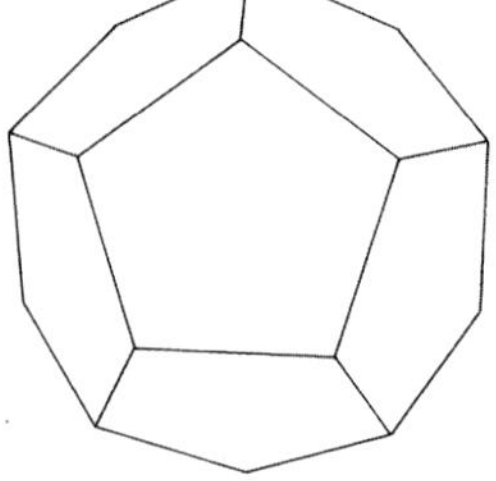 Dodecahedron	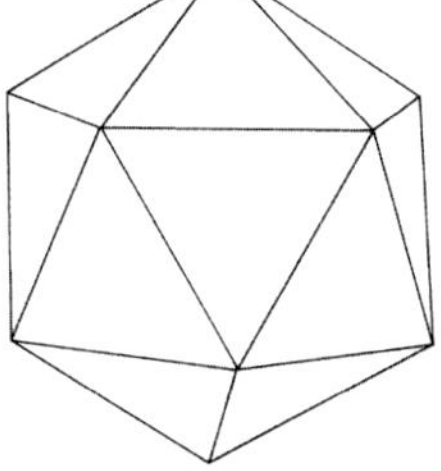 Icosahedron	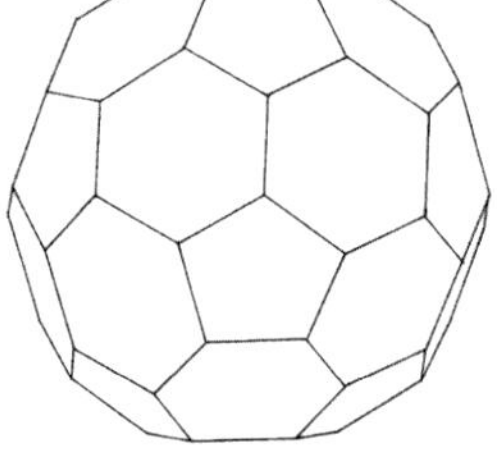 Truncated Icosahedron	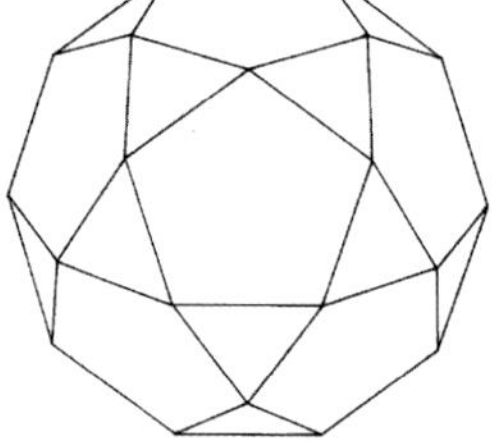 Icosidodecahedron

The symmetries of the polyhedra can be found by observing how many part turns would return the figure to its starting position. Each polyhedron has three different symmetries depending on whether you are looking at it face on, edge on, or vertex on – apart from the Tetrahedron and Truncated Tetrahedron whose face and vertex symmetries are the same. (See Appendix A for a further explanation of Symmetry. See also Keith Critchlow's *Order in Space* for a detailed 'Periodic Table' of the Platonic and Archimedean solids.)

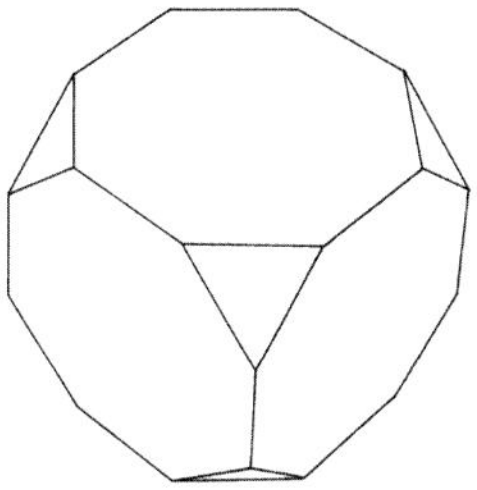 Truncated Cube	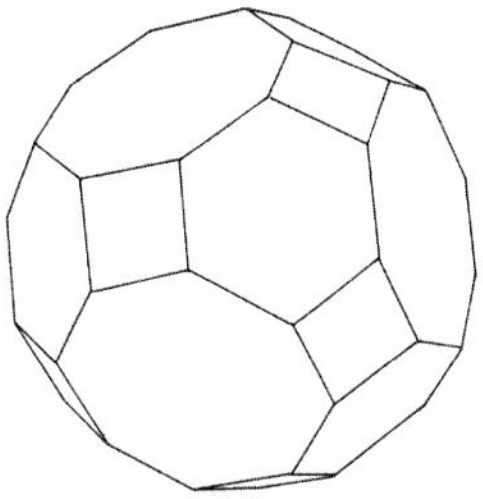 Truncated Cuboctahedron	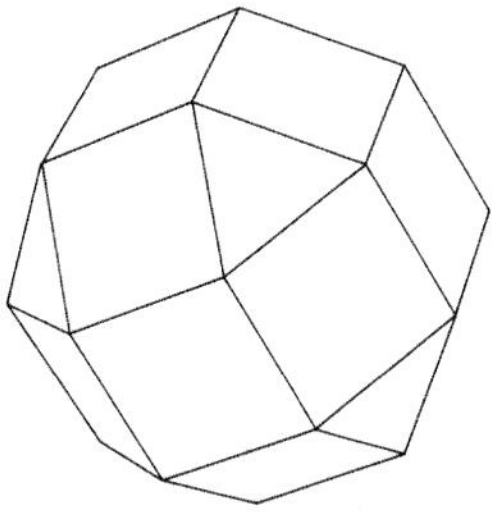 Rhombicuboctahedron	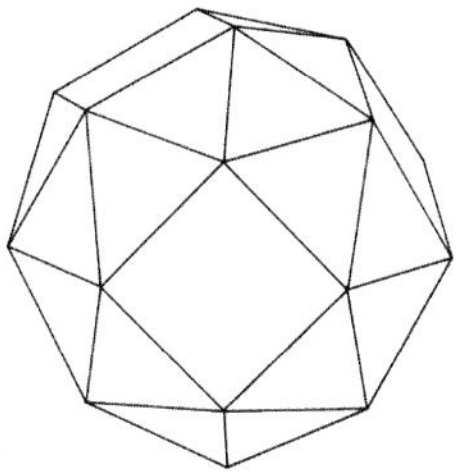 Snub Cube
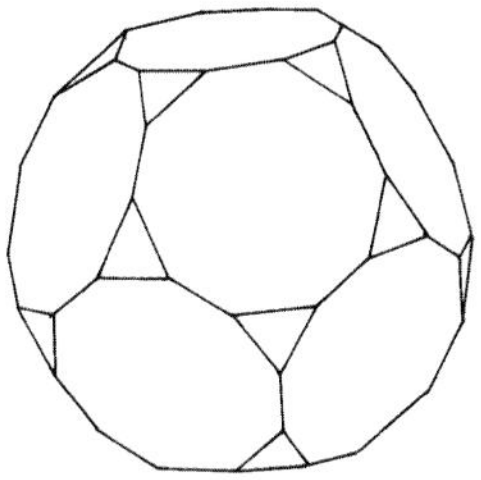 Truncated Dodecahedron	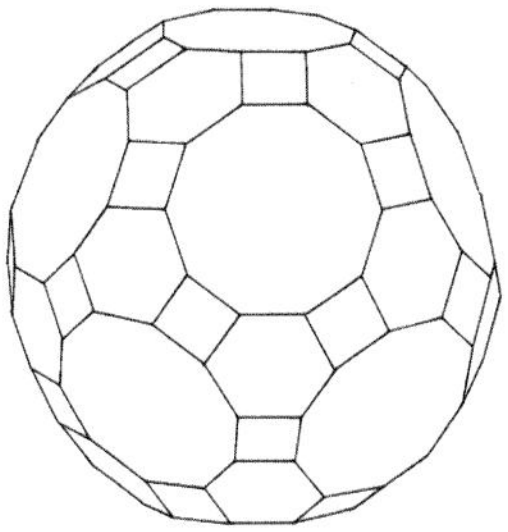 Truncated Icosidodecahedron	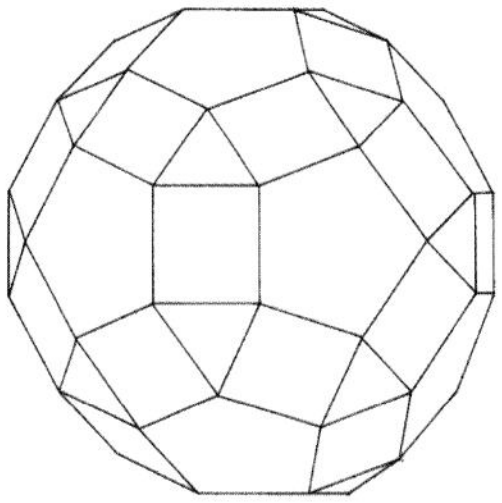 Rhombicosidodecahedron	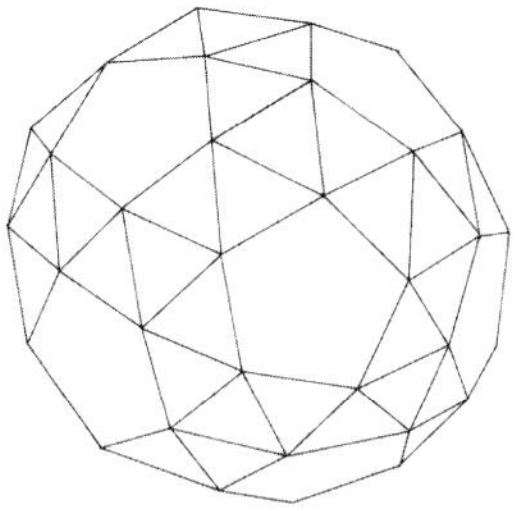 Snub Dodecahedron

Circles and Spherepoints

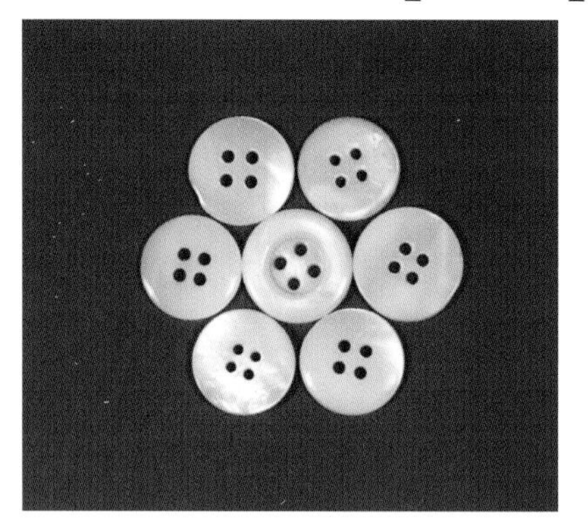

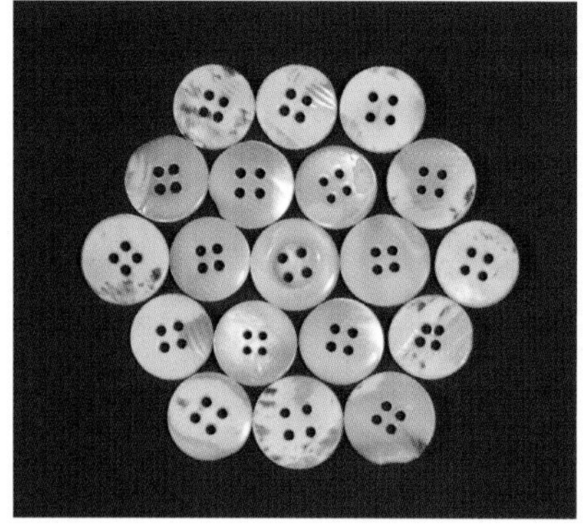

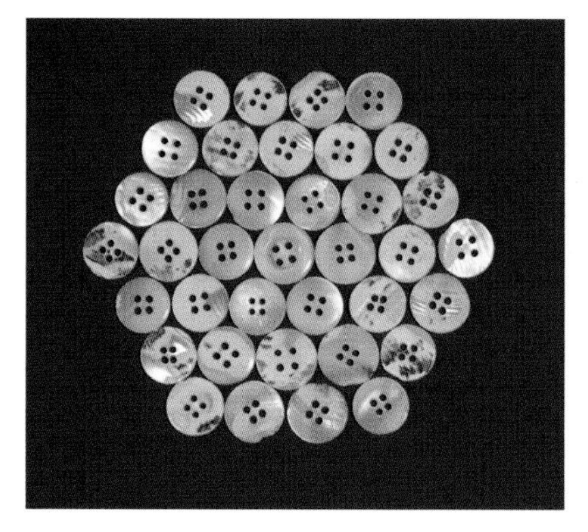

Figs. 1.1–1.3. Buttons arranged in one, two and three concentric hexagonal 'rings' about a centre.

The fundamental hexagonal relationship of circles in close packing can be demonstrated by taking circular objects of the same size (coins or buttons as here, for example) and surrounding a central circle with further rings. The first ring is composed of six circles, and each succeeding ring has a further six circles, so the sequence of circles in each ring is 1, 6, 12, 18, and so on, giving totals of 1, 7, 13, 19, 37, and so on. The unfolding hexagonal pattern can be seen in the illustrations above.

Table tennis balls can be used to demonstrate the build up of solid shapes. On the facing page top left we see the Tetrahedron; next to it the Octahedron; and bottom right the Icosahedron. The Icosahedron is composed of twelve spheres all touching five others, with a void in the centre which would be filled by a sphere of diameter about nine tenths that of the surrounding spheres. Twelve spheres can be fitted exactly around a central sphere of the same size, but in this case they take the form of the Cuboctahedron (Fig. 1.7 bottom left). You can see that the balls here make squares and triangles – in contrast to the Icosahedron where all the relationships are triangular.

Figs. 1.4–1.7. Tetrahedron, Octahedron, Cuboctahedron and Icosahedron made in table tennis balls.

Constructing Small 3D Stick Models

The Five Platonic Solids

Before you start:

1. You will need: sticks, cutting board, cutting knife (scalpel or craft knife) and glue (I use Studio Gum rubber adhesive). I buy plain round wooden sticks 2 or 2.5 mm in diameter and either 150 or 200 mm long. They can be coloured using water colours. Some craft suppliers offer pre-coloured sticks (see Appendix D for a list of suppliers).

Figs. 2.1-2.3. Students making geometry at the Prince's Foundation 2010 Summer School. Photos: Richard Ivey.

2. It is important to understand how the glue works. Studio Gum works well because it remains flexible for some time after it sticks, so models can be adjusted without falling to pieces. But first the tips of the sticks need to be dipped into the glue and set aside to become tacky – which means you should be able to hold one stick, attach another to the end, and dangle it without it falling off. If it falls off, the glue has not reached the required tackiness. If you try to stick your model together too soon, the glue will not do its job. It is best to work out how many sticks you will need for the particular part of the model you are making, glue up the required sticks, and then set them aside to become tacky. Apply glue to *both* ends to be joined together.

Tetrahedron

4 triangular faces, 4 vertices, 6 edges.

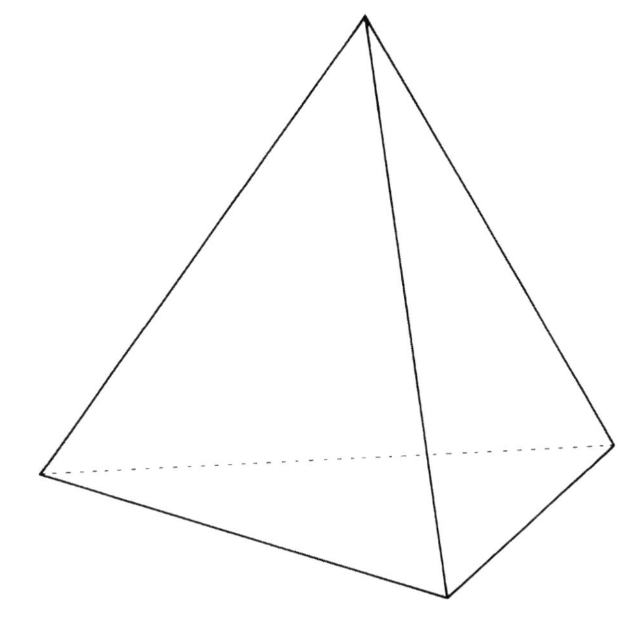

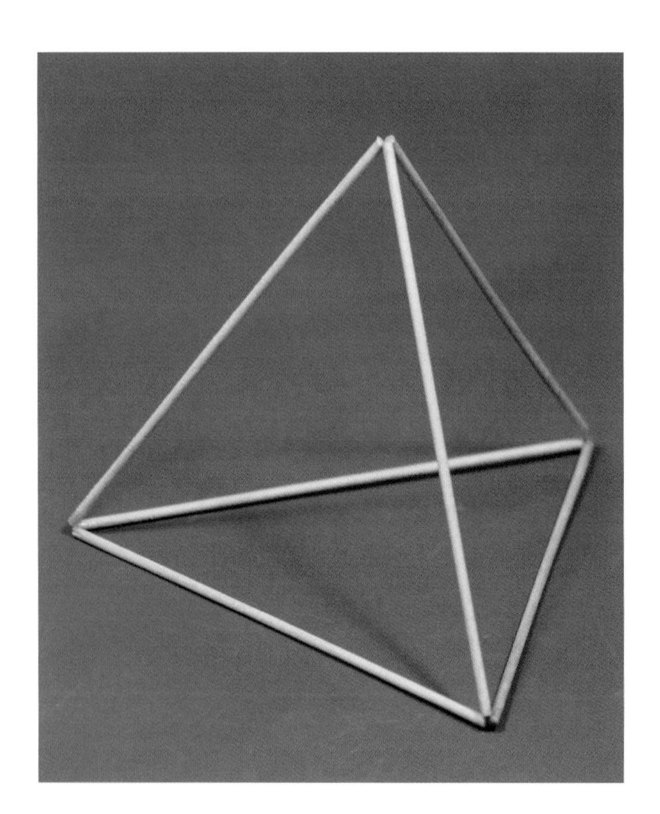

Fig. 2.4.

Method:

Required: 6 sticks (all one length)

1. Make a base triangle using 3 sticks.

2. Erect a tripod above (3 sticks brought together at the apex).

Cube (Hexahedron)

6 square faces, 8 vertices, 12 edges.

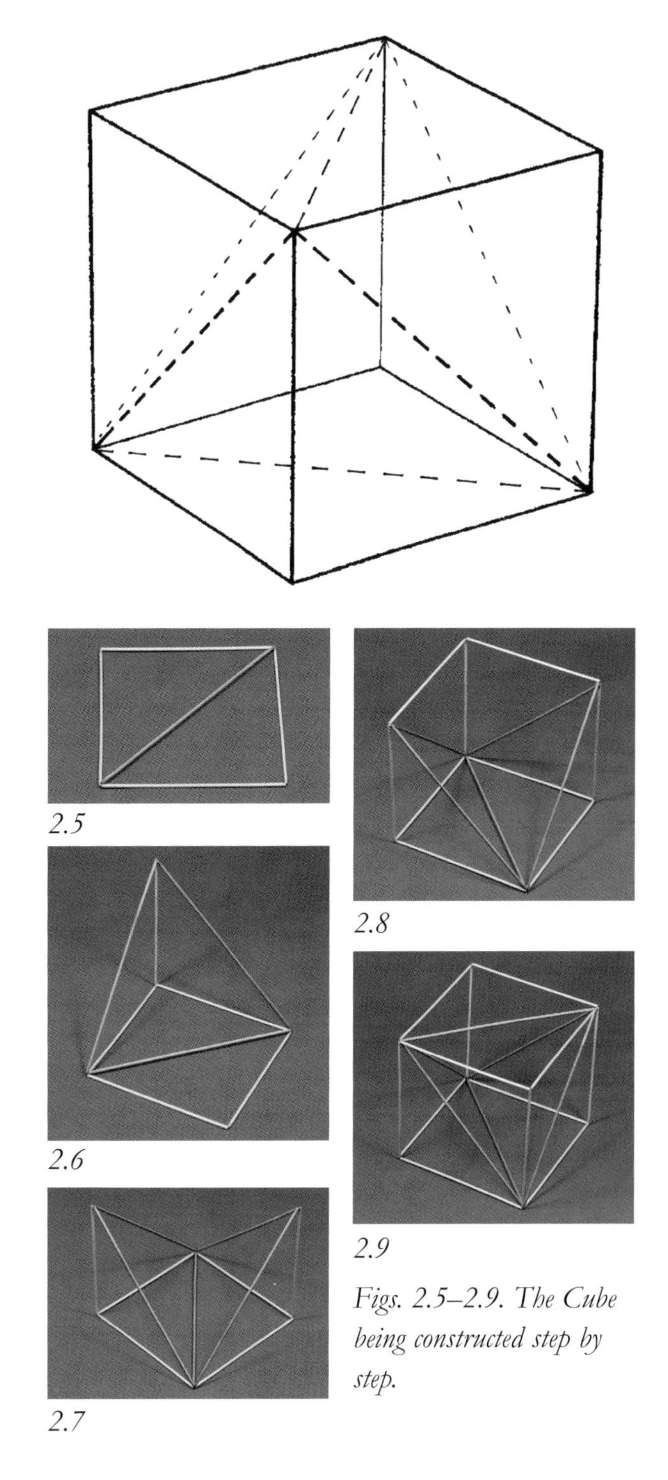

Method:

Required: 12 sticks (all one length); 6 sticks (all √2 or 1.414 times longer). Measure the short length, then either construct geometrically or use a calculator to work out the longer length.

1. Make a base square reinforced with one √2 diagonal (Fig. 2.5).

2. Erect one corner opposite the base diagonal braced by two diagonals (Fig. 2.6).

3. Erect the opposite corner braced again by two diagonals (Fig. 2.7).

4. Complete in turn each remaining corner, comprising one corner post and two sides of the top square (3 sticks each corner). Then add the final diagonal along the top (Figs. 2.8, 2.9).

Note that the diagonals are in the form of a Tetrahedron.

2.5

2.6

2.7

2.8

2.9

Figs. 2.5–2.9. The Cube being constructed step by step.

Octahedron

8 triangular faces, 6 vertices, 12 edges.

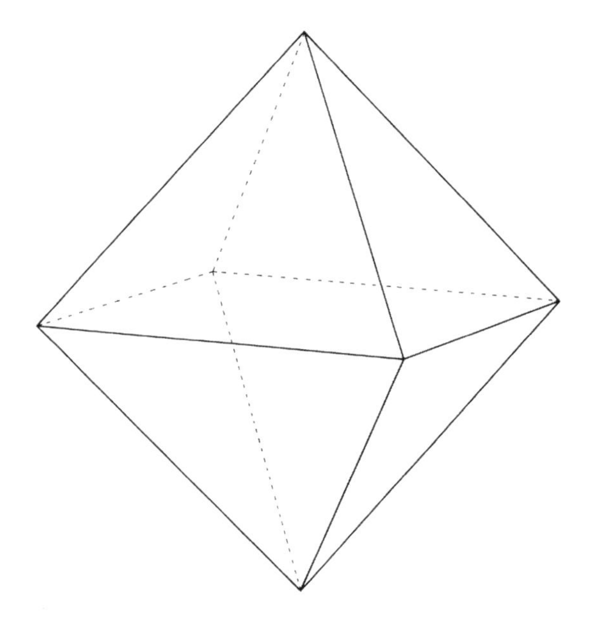

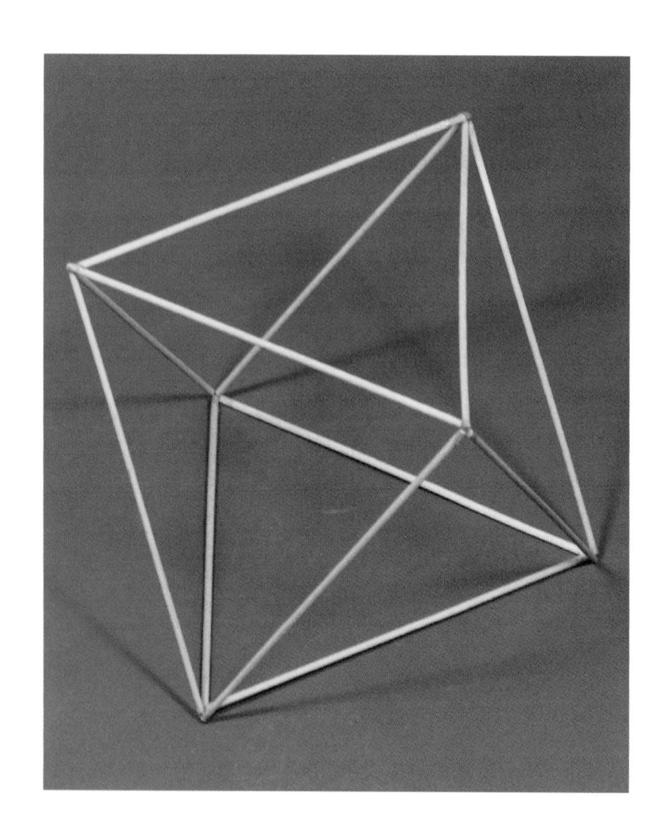

Fig. 2.10.

Method:

Required: 12 sticks (all one length); a second pair of hands helps.

1. Make FOUR separate triangles.

2. Lay one triangle down, attach one apex of each of the remaining three triangles to a different corner of the base triangle, and bring the free corners together.

Icosahedron

20 triangular faces, 12 vertices, 30 edges.

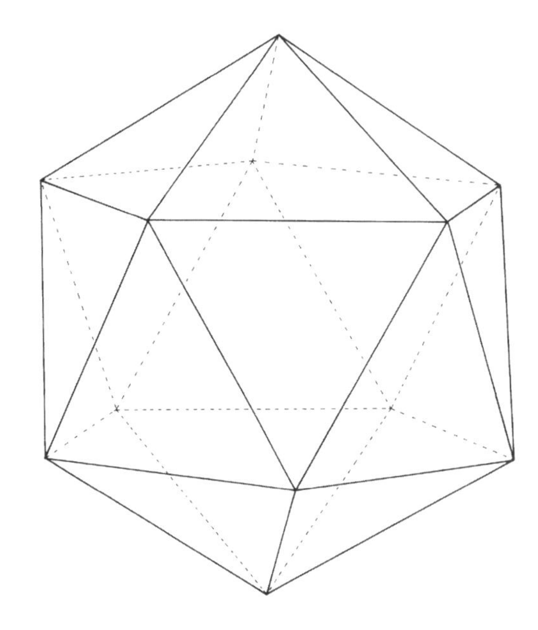

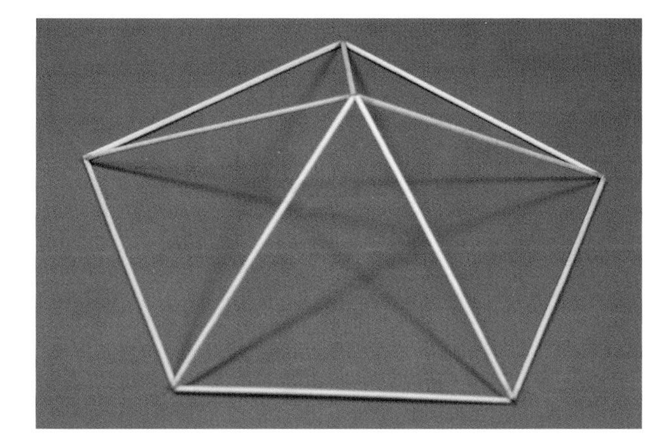

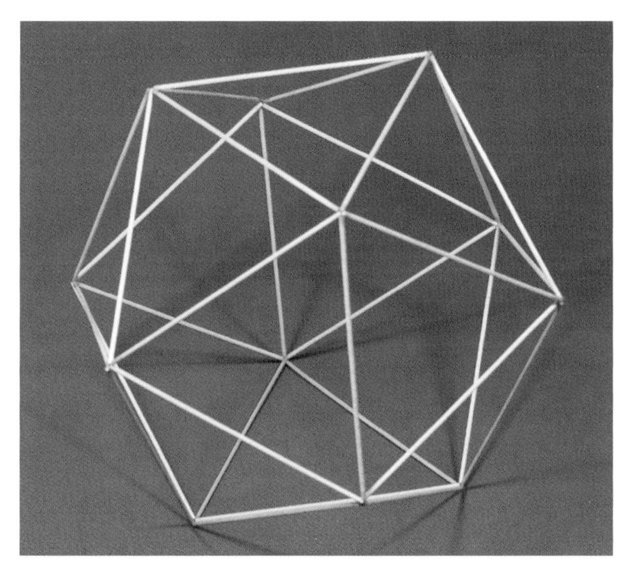

Figs. 2.11–2.12. The construction of an Icosahedron – step 2 (top) and complete (bottom).

Method:

Required: 30 sticks (all one length); a second pair of hands is very helpful.

1. Take 10 sticks and create two (rough) pentagons.

2. For each pentagon, take 5 more sticks and, sticking one end of each stick to a corner of the pentagon, draw all five together to an apex (Fig. 2.11).

3. Holding one shape up, hang triangles down from each side of the pentagon using 2 further sticks each time.

4. From below, offer up the second pentagonal shape (one corner to each dangling triangle) to complete the Icosahedron.

Dodecahedron

12 pentagonal faces, 20 vertices, 30 edges.

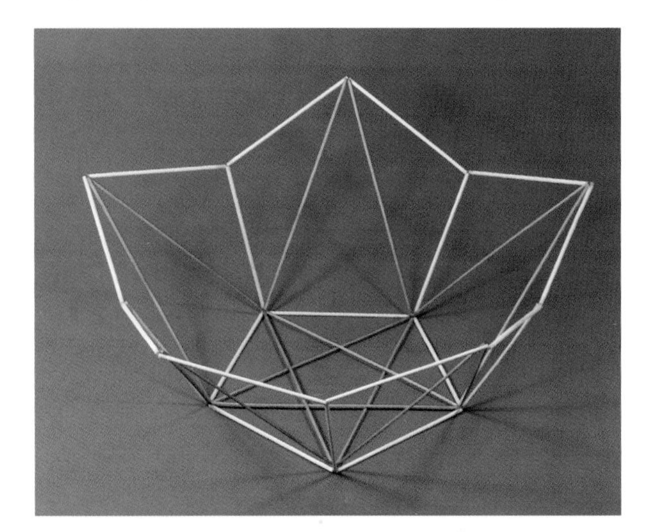

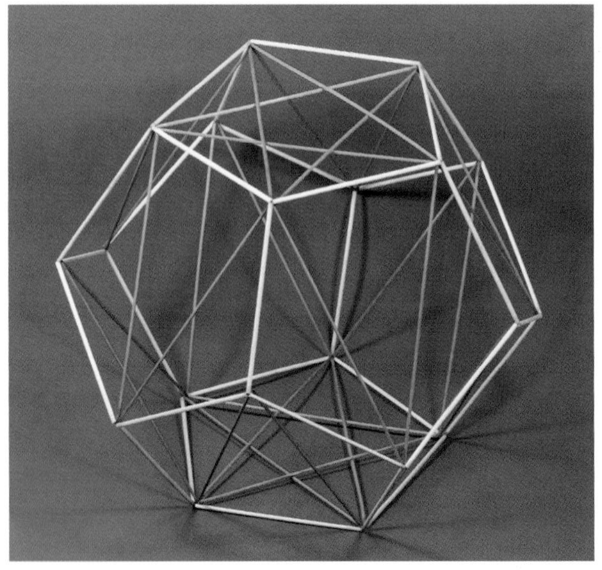

Figs. 2.13–2.15. Illustrations of one base pentagon plus the second base pentagon with reinforcing triangle added (facing page); the base half of the model made (this page above); and the figure complete (below).

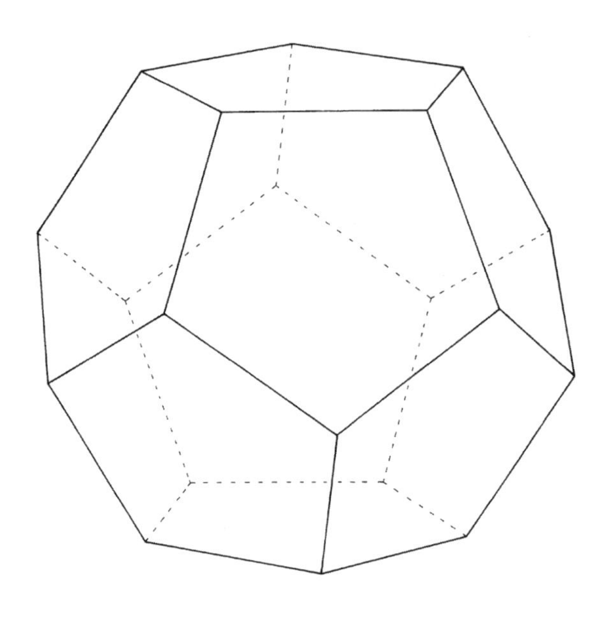

Method:

Required: 30 sticks (all one length); 30 sticks (all Φ or 1.618 times longer); a second pair of hands is essential.

1. Create 2 base pentagons from 10 short sticks, each reinforced with 5 long sticks in the shape of a pentagram (bottom of Fig. 2.13 on facing page).

2. To each side of both base pentagons, add a pair of long sticks to create a triangle (20 sticks all told). You will then have two large

pentagrams (top of Fig. 2.13 this page).

3. With a little help from a friend, take one base pentagram and add 3 short sticks in the shape of a 'T' to each corner, gathering up the reinforcing triangle of longer sticks to each apex. When done, the 'dish' will be wobbly but will hold itself by leaning out under its own weight (Fig. 2.14).

4. While your helper holds the second base pentagon, carefully join the apex of each dangling triangle into the lower five points of the edge of the 'dish' you have just made. When done, the model will again hold itself.

5. Finally, add the remaining 5 short sticks to complete the figure, which now becomes fully stable.

Further Demonstrations

The stick models can be used very effectively to demonstrate some of the remarkable characteristics of the regular polyhedra. The five Platonic solids nest within each other in any and every order. Six of these nestings can be demonstrated using the stick models by tying string from one point to another. Three relationships are vertex to vertex – such as the Tetrahedron within the Cube (Fig. 3.1 below left); and three are vertex to side midpoint – such as the Octahedron within the Tetrahedron (Fig. 3.2 below right).

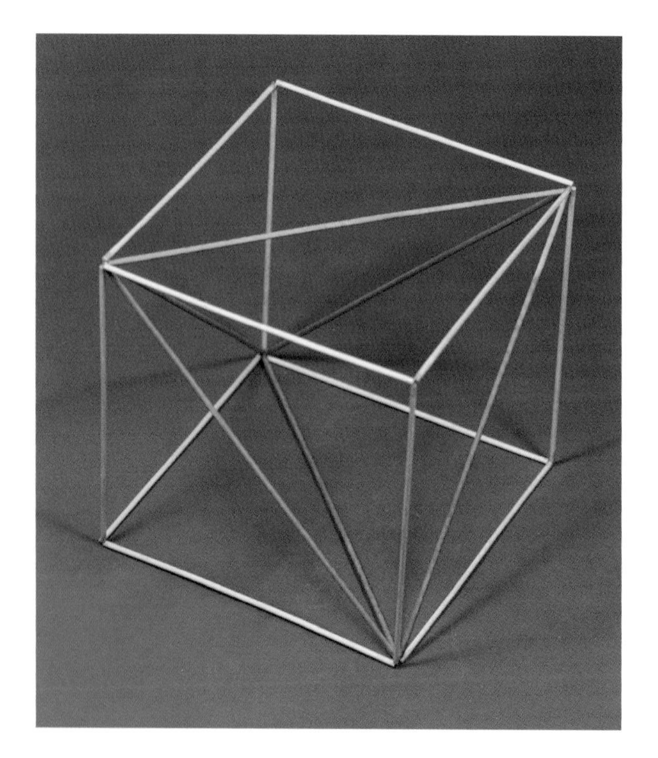

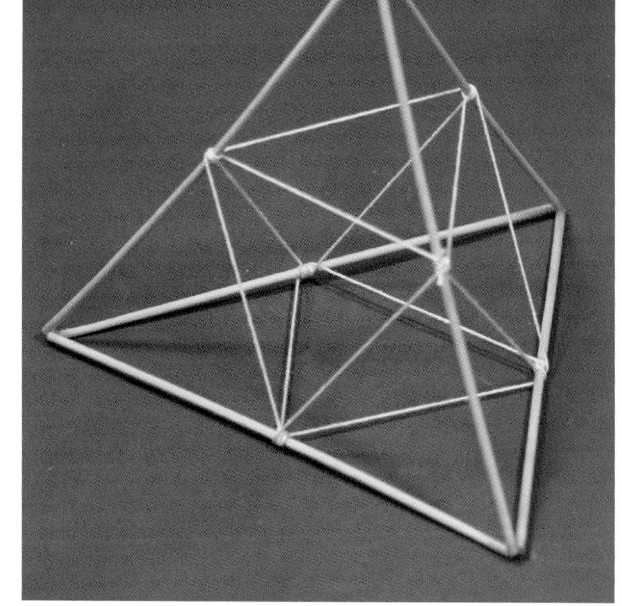

Figs. 3.1–3.2. Tetrahedron within a Cube (left); Octahedron within a Tetrahedron (right).

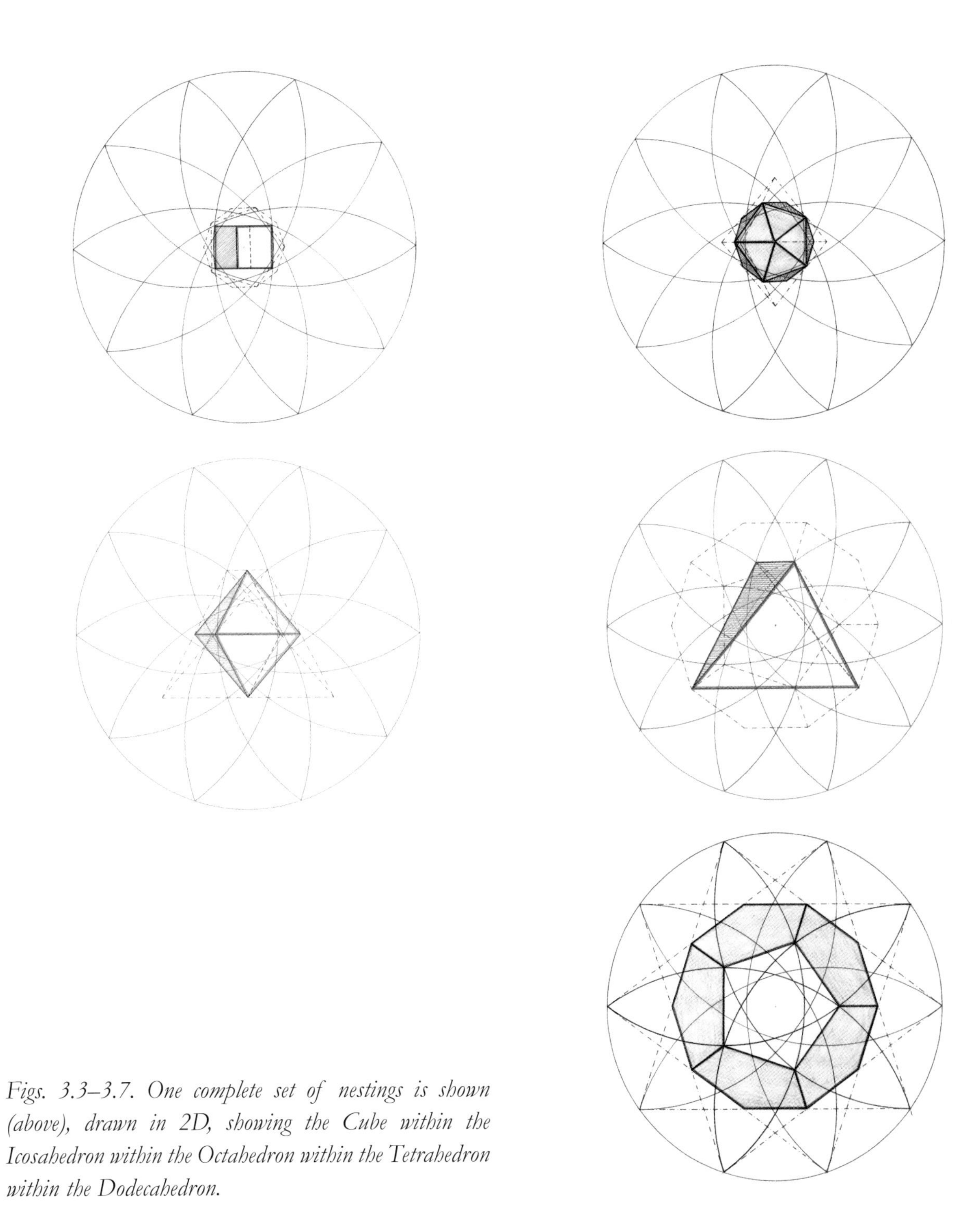

Figs. 3.3–3.7. One complete set of nestings is shown (above), drawn in 2D, showing the Cube within the Icosahedron within the Octahedron within the Tetrahedron within the Dodecahedron.

Other nesting relationships can be found, such as the Cuboctahedron within the Octahedron (Fig. 3.8 below).

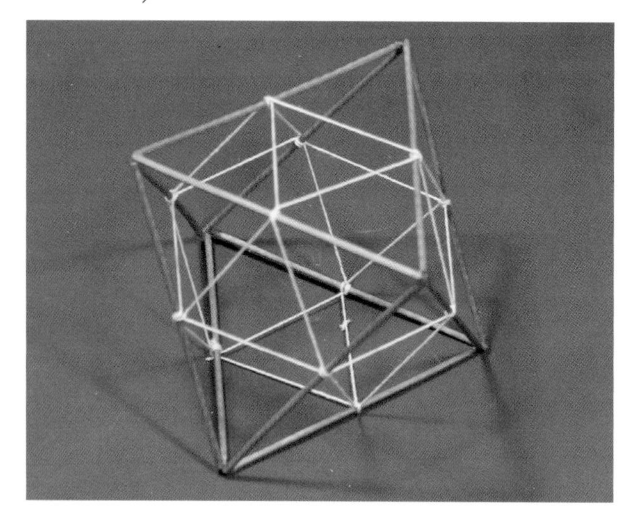

Fig. 3.8. Cuboctahedron within an Octahedron.

The stick models can also be 'dressed' in detail paper (or Tervakovsky paper if you can find it) to create 'solid' models by cutting paper to size and gluing to the sticks; or cutting oversize and trimming to size once glued.

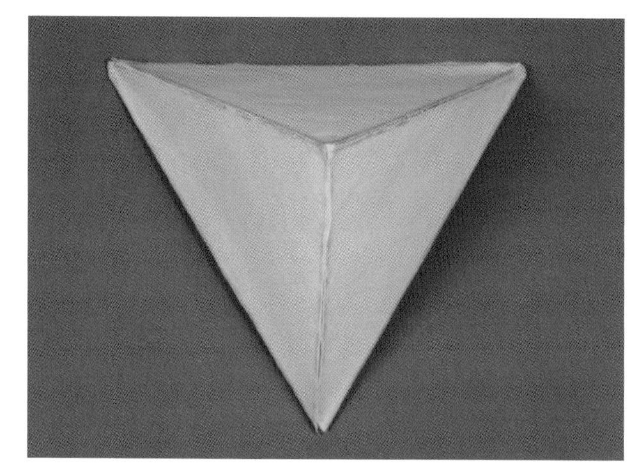

Figs. 3.9–3.11. Tetrahedron, Octahedron and Icosahedron in paper on stick frames.

One further use can be made of the stick models, to show the internal rectangles which in many cases contain a special geometric proportion. The fifteen internal rectangles within the Icosahedron, for example, are all golden (Φ) rectangles (one such rectangle is picked out in Fig. 3.12 above right). The rectangles within the Cube are all √2 rectangles (Fig. 3.13 below right).

Fig. 3.12. The paper sheet shows one of the fifteen internal golden rectangles within the Icosahedron.

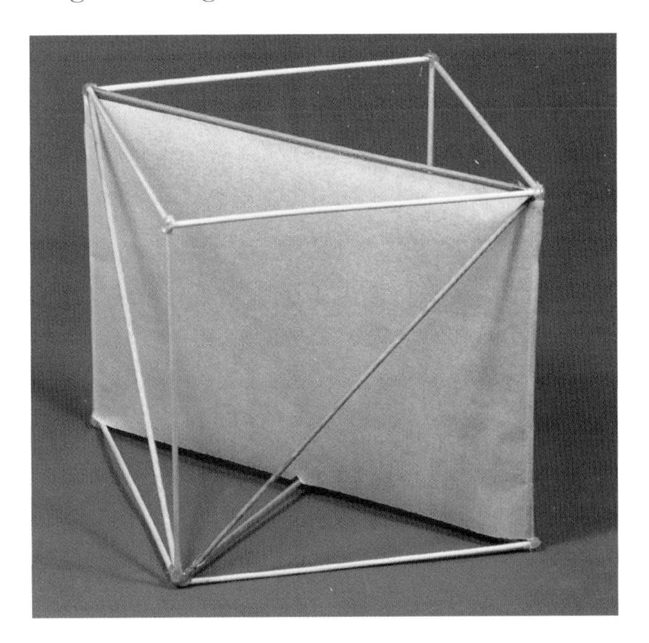

Fig. 3.13. The paper sheet shows one of the six internal √2 rectangles within the Cube.

Paperfolds for the Five Platonic Solids

Fig. 4.1. The Five Platonic Solids made in card.

Before you start:

1. A suitable weight of paper depends on the size of model that you are making. 300g/m^2 (watercolour or coloured pastel paper) gives you a very stout model, but achieving neat creases can be difficult. Probably best is heavyweight cartridge paper (200 or 220 g/m^2), which makes good models, can be nicely creased, and can be painted. You could even use 160 g/m^2 (suitable for use in photocopiers) as long as the model is not so large that it becomes flimsy. I usually use a paper size no bigger than A3 to draw out the nets – using several sheets where the net is drawn out in multiple parts.

2. In addition to your straight edge, pencil and compasses, you will find it useful to have two coloured pencils to distinguish between lines to be cut and lines to be scored and folded. If you do not have coloured pencils, distinguish between the two types of lines by using bold solid lines for the 'cut' lines, and broken (dashed) lines for the 'score and fold' lines (as used here throughout in the instruction sheets).

3. You will also need a scalpel or craft knife, a cutting board, and glue (Uhu General Purpose is good).

4. Draw the construction faintly at first, then strengthen the 'cut' and 'fold' lines when you have completed the whole drawing. The finished model looks best if you fold your card so that the score lines as well as the drawn construction are inside, and gives you clean surfaces for decoration.

5. If you are not experienced in scoring, practise first to work out how much pressure is needed for the scored line to fold nicely. Score too deeply and your tabs will fall off! With practice, the back of a scalpel blade works well with thick watercolour paper; and the back of an old blunt knife works better on thinner card.

Net for the Tetrahedron (The Platonic Solid of Fire)

4 triangular faces, 4 vertices, 6 edges. Tetrahedral (2, 3-fold) symmetry.

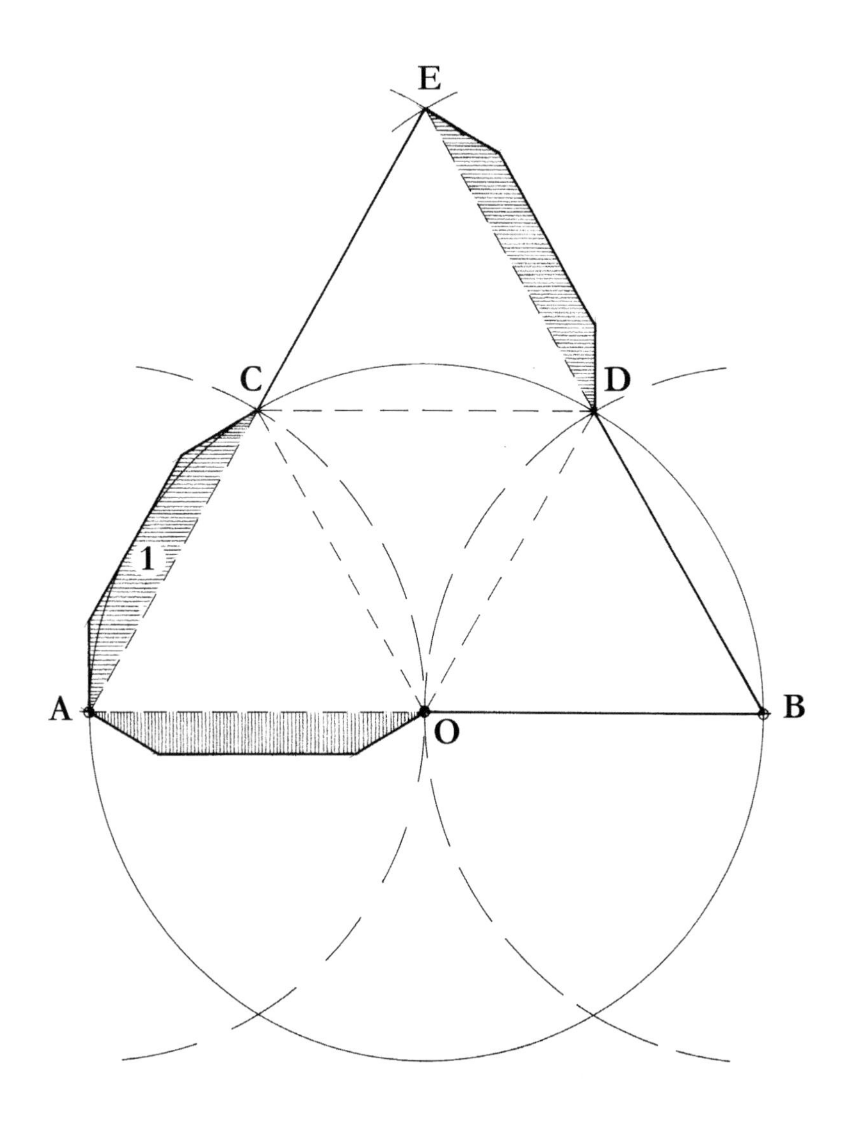

Fig. 4.2. Construction drawing.

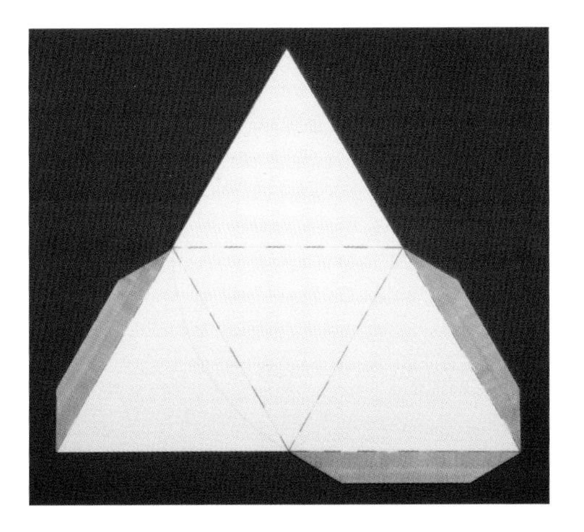

Figs. 4.3–4.5. Illustrations of the net drawn out on card and cut out (left); part assembled (centre); and the model complete (right).

Method:

1. Draw a circle centre O – (judge the position on the sheet, allowing space for top point E).

2. By eye, draw a long horizontal through the centre of the circle parallel to the bottom edge of the sheet.

3. Draw 2 arcs with the same radius from the diameter points A and B to intersect the circle above and below. The intersections above will give you points C and D.

4. To find E: either, use your straight edge to extend AC and BD to cross; or, draw arcs from C and D to cross.

5. Add tabs as shown.

6. Cut out along the solid lines: score along the dashed lines.

7. Fold upwards along the scored lines.

8. Glue carefully (tabs inside) starting with tab marked 1.

9. Clean up and decorate as you wish.

Net for the Octahedron (The Platonic Solid of Air)

8 triangular faces, 6 vertices, 12 edges. Octahedral (2, 3, 4-fold) symmetry.

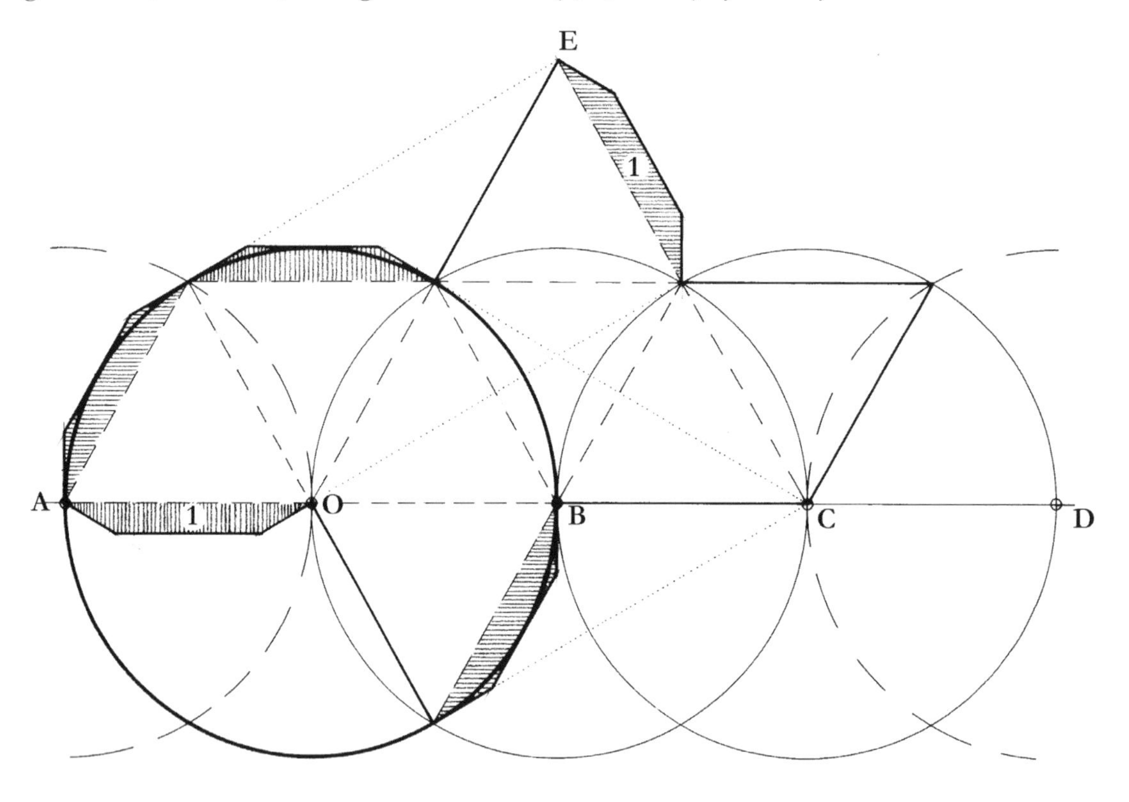

Fig. 4.6. Construction drawing.

Fig. 4.7. Illustration of the net drawn out on card and cut out.

Figs. 4.8–4.10. Illustrations of the net in two stages of assembly, and the model complete (right).

Method:

1. Draw a circle with centre O – (judge the position on the sheet as shown, to allow space for top point E).

2. Draw a long horizontal through the centre of the circle parallel to the bottom edge of the sheet.

3. Along the horizontal, with the same radius, draw 2 more circles with centres B and C, and 2 semicircular arcs (shown dashed) centres A and D.

4. Using circle intersection points, draw out the net as shown on facing page (Fig. 4.6).

5. Add tabs as shown: the cut angle can be obtained with reference to points on the drawing as shown. Judge the width by eye.

6. Cut out along the solid lines: score along the dashed lines. Fold upwards along the scored lines. Glue carefully (tabs inside) starting with tabs marked 1. Clean up and decorate as you wish.

Net for the Icosahedron (The Platonic Solid of Water)

20 triangular faces, 12 vertices, 30 edges. Icosahedral (2, 3, 5-fold) symmetry.

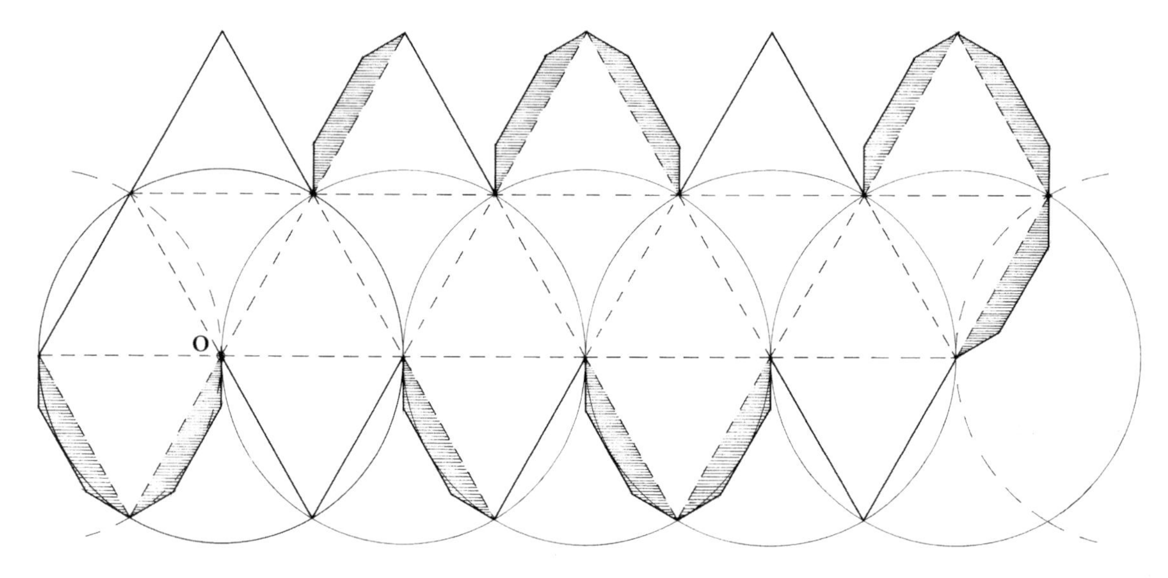

Fig. 4.11. Construction drawing.

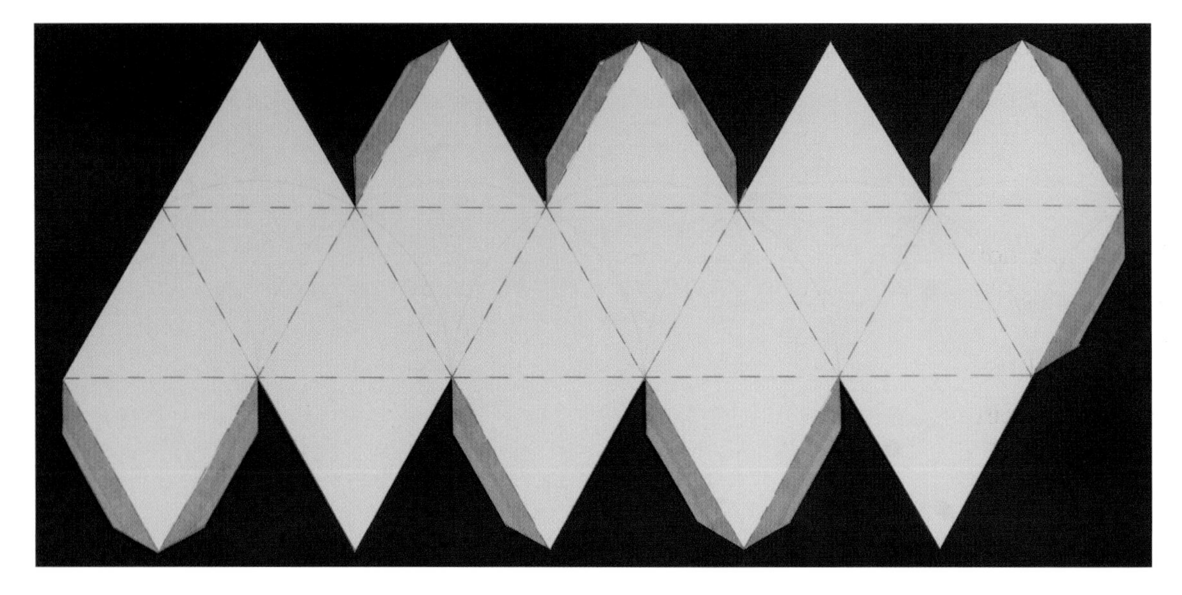

Fig. 4.12. Illustration of the net drawn out on card and cut out.

Figs. 4.13–4.14. Illustrations of the model partially assembled (left) and complete (right).

Method:

1. Draw a circle centre O just in from the edge of the sheet and with a radius no more than one sixth of the width of your sheet (see Figure 4.11).

2. Draw a long horizontal through the centre of the circle parallel to the bottom edge of the sheet.

3. Draw 4 more circles along the horizontal overlapping by half a diameter each time. Draw semicircular arcs at the extremes to left and right (shown dashed).

4. Using circle intersection points draw out the net as shown above.

5. Add tabs as shown using points on the drawing as guides to angle them off.

6. Cut out along the solid lines: score along the dashed lines. Fold upwards along the scored lines. Glue carefully and progressively together (tabs inside). I find it easier to glue one or two tabs at a time, and then pause. Clean up and decorate as you wish.

Net for the Cube (The Platonic Solid of Earth)

6 square faces, 8 vertices, 12 edges. Octahedral (2, 3, 4-fold) symmetry.

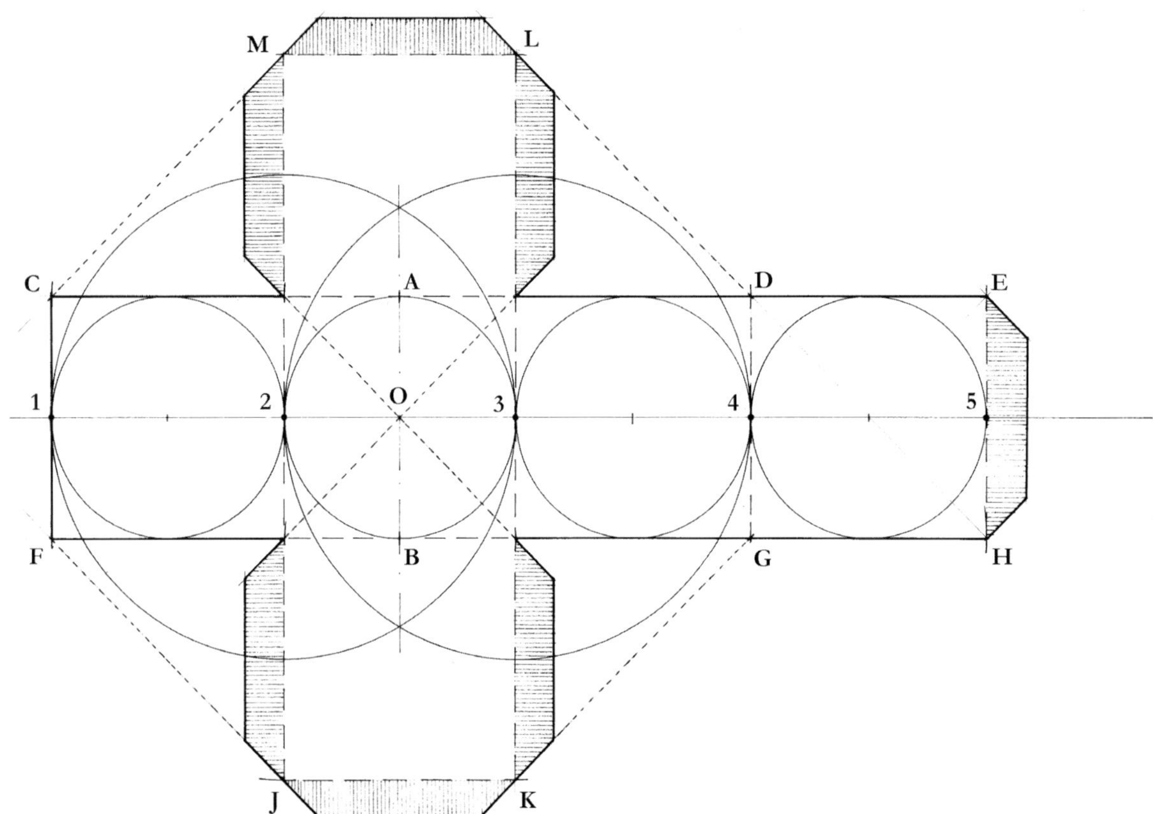

Fig. 4.15. Construction drawing.

Method:

1. Draw a horizontal in the centre of your sheet. Using your compass, mark off four equal measures (points 1, 2, 3, 4 and 5 in Figure 4.15 above) – with a bit to spare at each end; making sure also that there is room vertically for three measures plus something to spare top and bottom.

2. On points 2 and 3 draw circles with radius the 'measure'.

3. Draw the vertical through the two points where the circles intersect to give the midpoint O, as well as A and B.

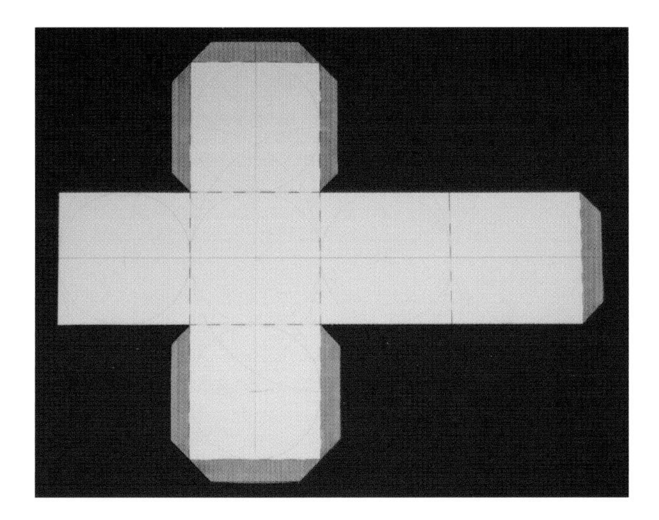

Fig. 4.16. Illustration of the net drawn out on card.

4. Draw a circle centre O radius O2. Repeat along the horizontal between points 1 and 2, 3 and 4, 4 and 5.

5. Draw horizontal lines tangenting above and below the four circles.

6. Mark off the corner points of the square about circle O using radius OA from points A and B. Connect the corner points and extend above and below.

7. Setting compass to the original measure, mark off corner points for the rest of the squares required (points C, D, E, F, G, H, J, K, L, M).

8. Add tabs at 45° as shown using diagonals where possible. Judge the width by eye.

9. Cut out along the solid lines: score along the dashed lines. Fold upwards along the scored lines.

10. Glue carefully and progressively (tabs inside). Clean up and decorate as you wish.

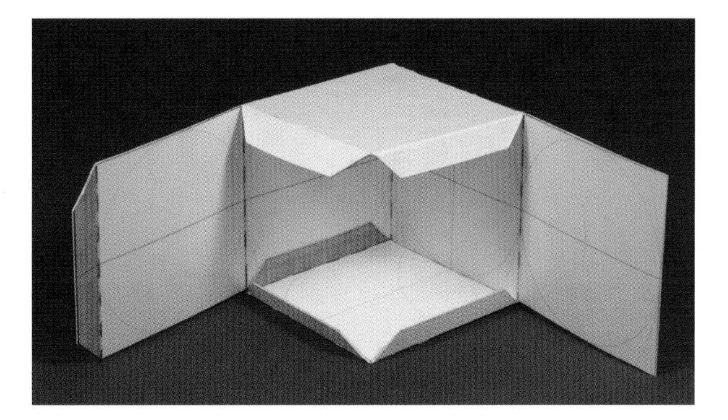

Figs. 4.17–4.18. Illustrations of the model partially assembled (above) and complete (below).

Net for the Dodecahedron (The Platonic Solid of Ether)

12 pentagonal faces, 20 vertices, 30 edges. Icosahedral (2, 3, 5-fold) symmetry.

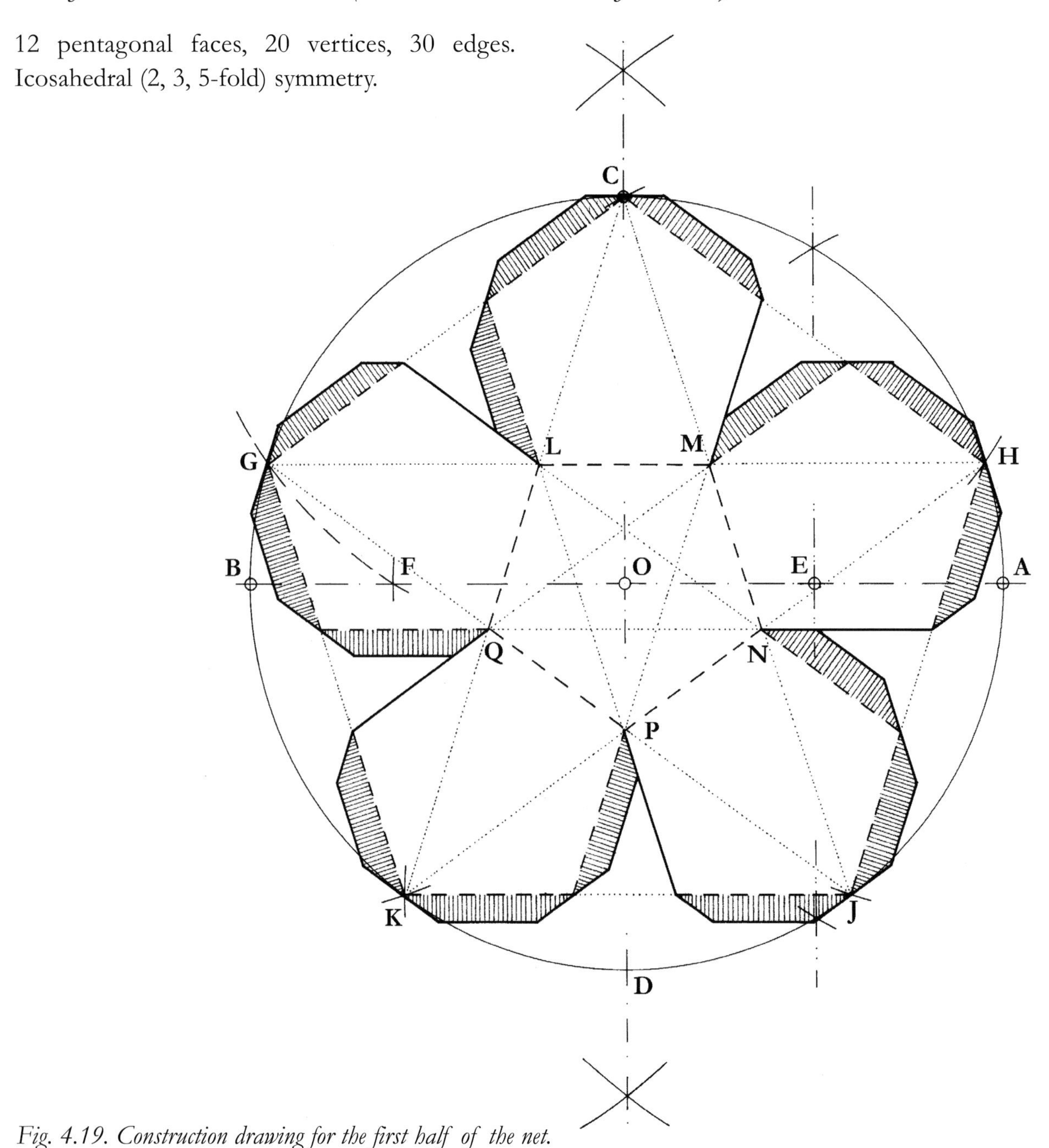

Fig. 4.19. Construction drawing for the first half of the net.

Method:

1. Draw a circle centre O radius OA, and the pentagon CHJKG within as follows: draw any horizontal diameter AB through O, and then the perpendicular vertical axis CD (equal arcs above and below from A and B). Find the mid-point E of OA (arc radius OA from A to cut circle above and below, and connect). Draw an arc centre E radius EC to cut BA at F: draw arc centre C radius CF to cut circle at G. CG is one side of our pentagon, which can be 'walked' around the circle with the compasses to find the other points H, J and K. Adjust if necessary so that the five points are equally spaced around the circle.

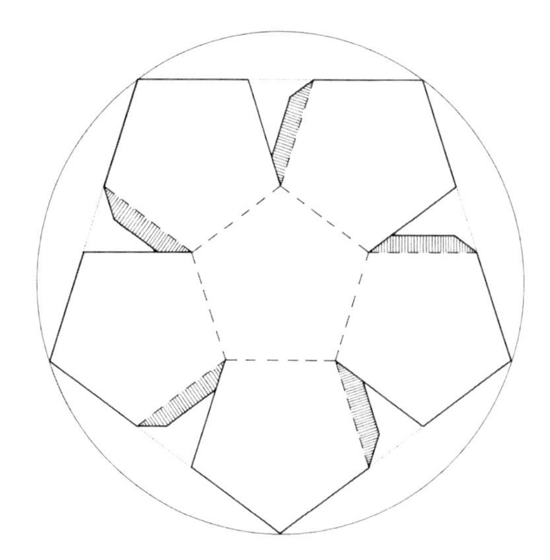

Fig. 4.20. Construction drawing for the second half of the net. Note that this must be of the same size as the first half.

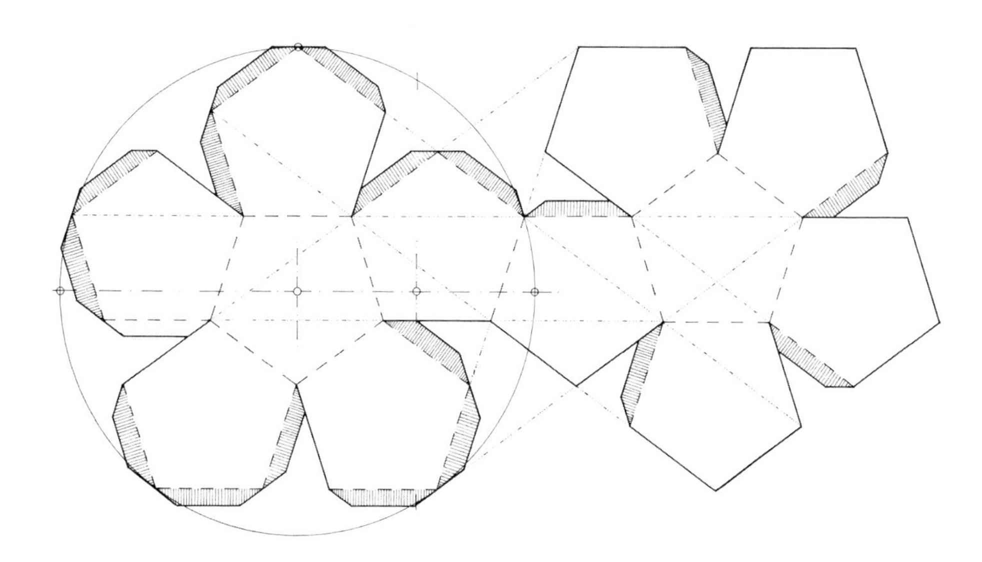

Fig. 4.21. With some care the whole net can be drawn out as one drawing. This requires some skill in the projection of lines from the first half (on the left within the circle) to ensure accuracy – which is why many will find it easier to make the net in two halves.

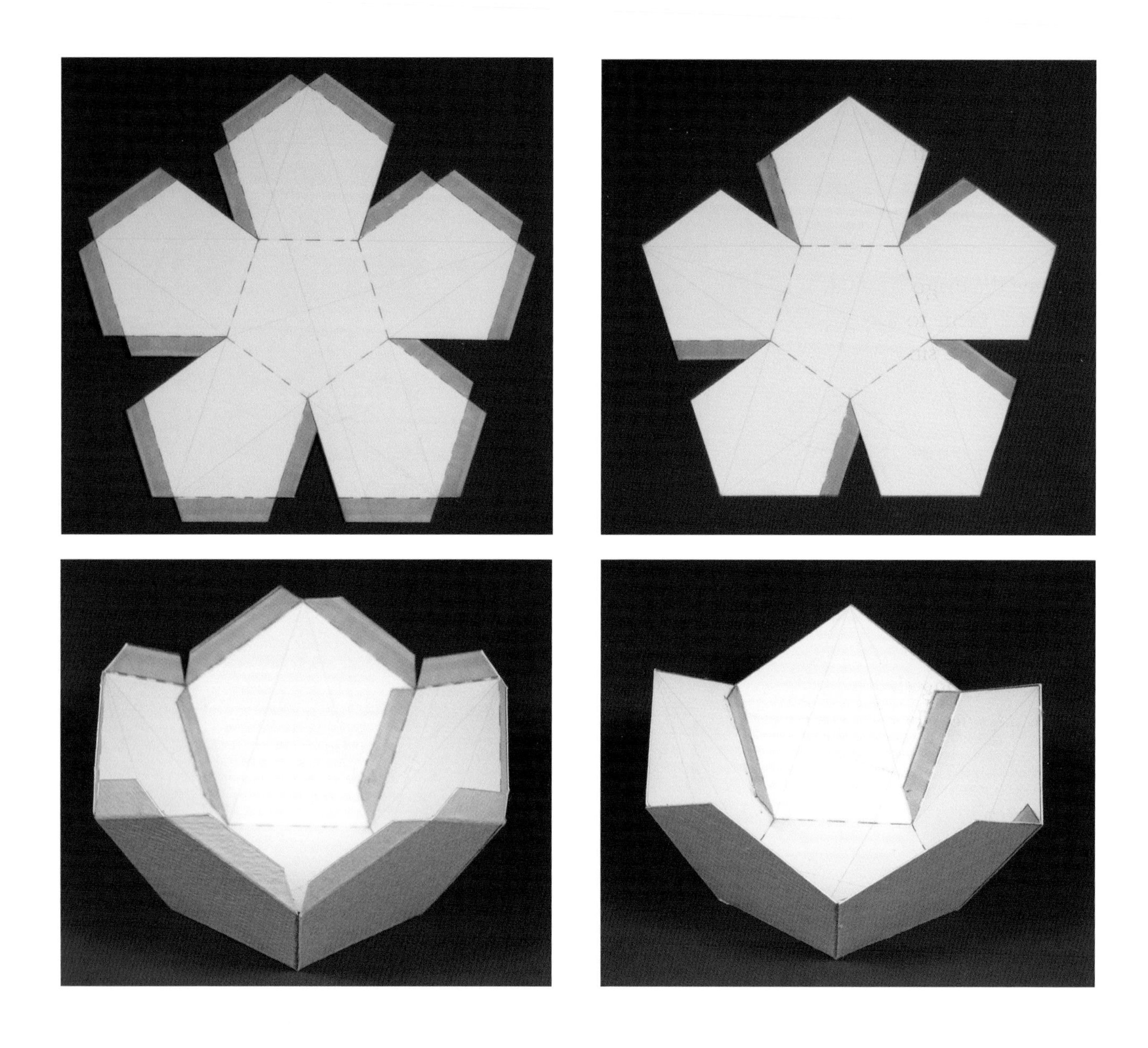

Figs. 4.22–4.25. The two halves of the net drawn out on card and cut out (top); and glued into dishes (bottom).

2. Draw a pentagram (star) within the pentagon by drawing lines from each point to all the other points (from G to H and J; from C to J and K; etc). In the middle of this pentagram is a smaller pentagon LMNPQ.

3. Draw the diagonals of this inner pentagon and extend them to intersect the main pentagon, creating five similar (same size) pentagons surrounding the inner one.

4. Mark the cut and fold lines as shown.

5. Add tabs as shown.

6. Make a second net of exactly the same size; but omit the outer tabs.

7. Cut out along the solid lines: score along the dashed lines. Fold upwards along the scored lines.

8. Glue both nets carefully and progressively (tabs inside) – one or two tabs at a time. Make sure the outer tabs are not folded too much as they need spring to engage with the second half. Then glue the outer tabs and join one half to the other.

9. Clean up and decorate as you wish.

Fig. 4.26. The two halves joined to complete the Dodecahedron.

Note:

See also Euclid's method for constructing a Dodecahedron on page 100.

Paperfolds for the Thirteen Archimedean Solids

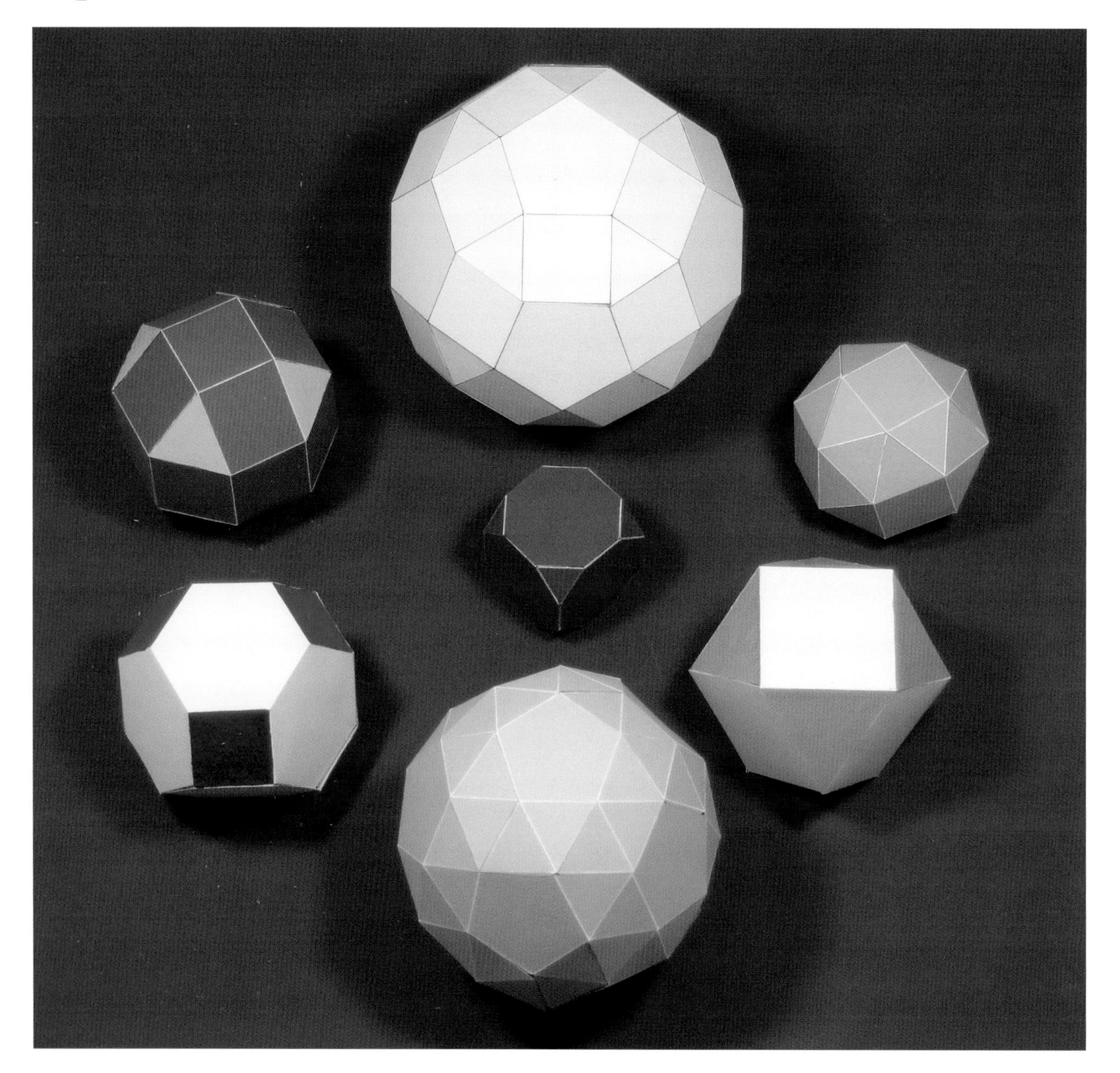

Fig. 5.1. A selection of the Archimedean solids – (clockwise from the top) – Rhombicosidodecahedron, Snub Cube, Cuboctahedron, Snub Dodecahedron, Truncated Octahedron, Rhombicuboctahedron, and (in the centre) Truncated Cube.

Net for the Truncated Tetrahedron

8 faces (4 triangles, 4 hexagons), 12 vertices, 18 edges. Tetrahedral (2, 3-fold) symmetry.

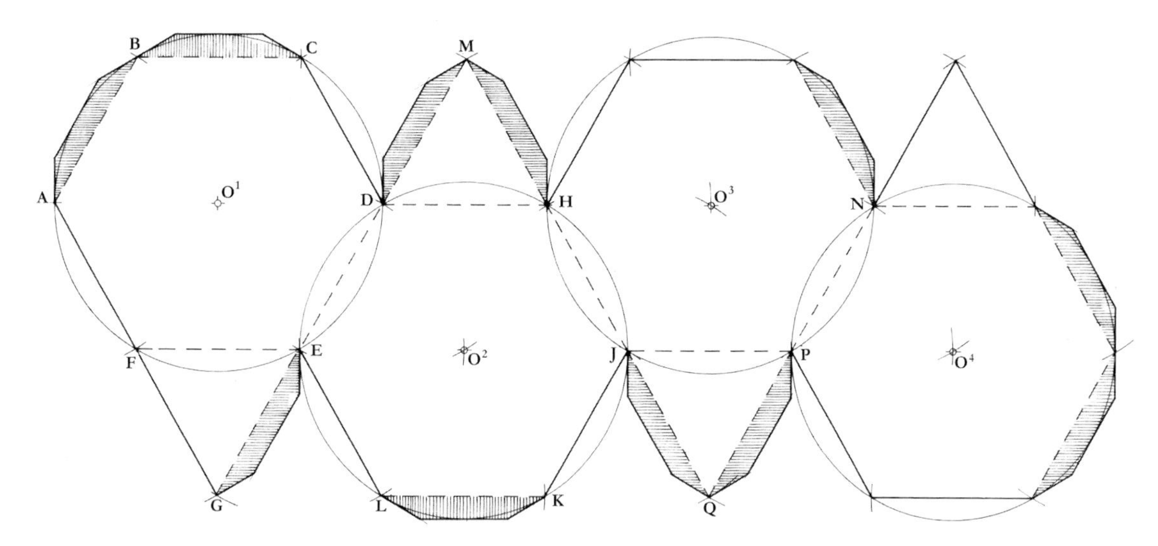

Fig. 5.2. Construction drawing.

Method:

1. The whole construction is drawn using the same radius throughout. Draw a circle centre O^1 radius O^1A. Mark off points B, C, D, E and F on the circle by walking the radius round. You should come back exactly to A from F. Draw arcs from E and F to find G; and from D and E to find second circle centre O^2.

2. Draw second circle from O^2 which will pass through points D and E. Walk the radius round the circle to find points H, J, K and L. Check that you return exactly to E from L. Draw arcs from D and H to find M; and from H and J to find third circle centre O^3.

3. Draw third circle centred on O^3, and walk radius round as before. Draw arcs as before to find Q, and fourth centre O^4.

4. Follow the same process for the fourth circle to complete the construction.

5. Add tabs and mark the cut and fold lines as shown. Cut out along the solid lines: score along the dashed lines. Fold upwards along the scored lines. Glue carefully and progressively (tabs inside) – one or two tabs at a time. Clean up and decorate as you wish.

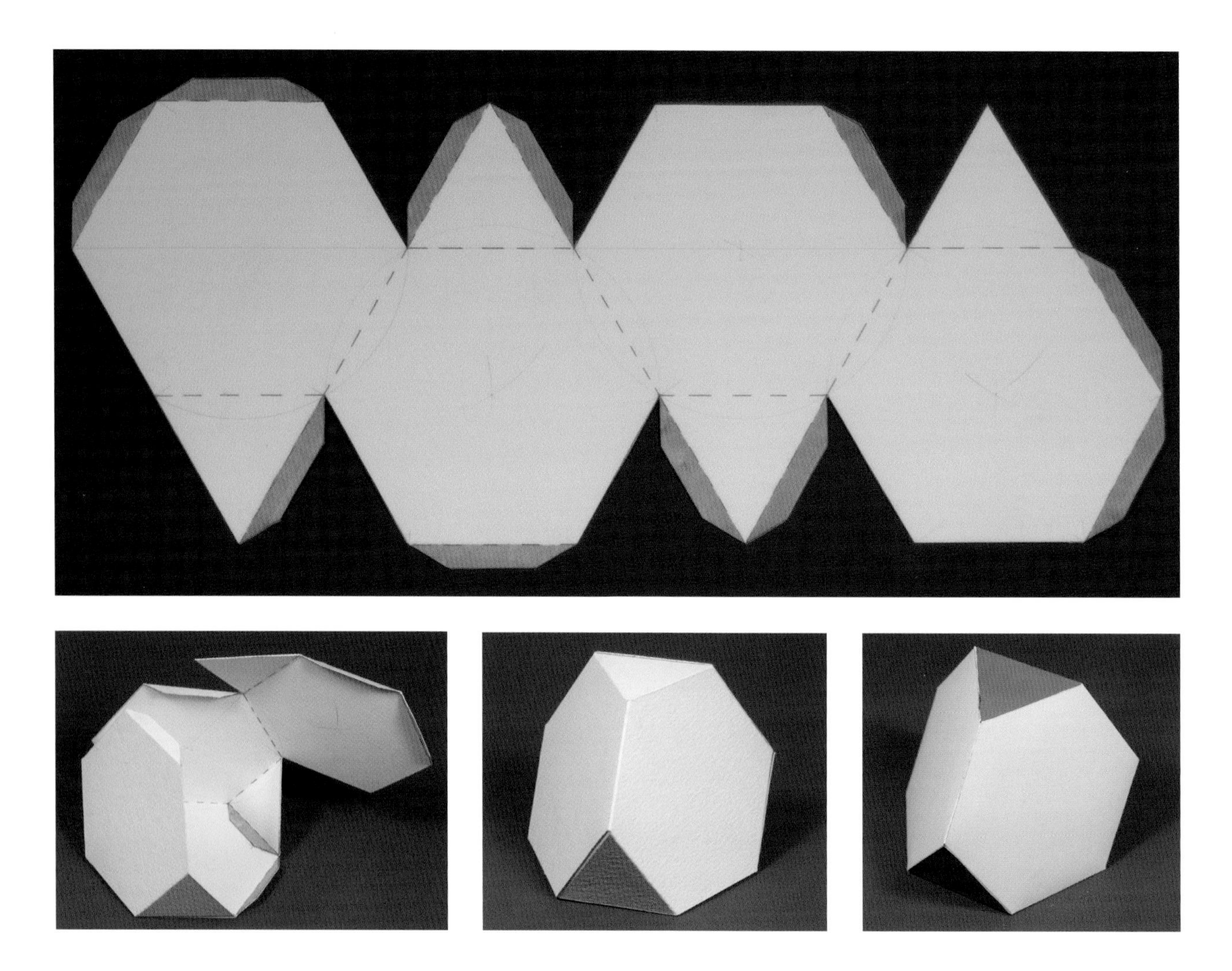

Figs. 5.3–5.6. Illustrations of the net drawn out on card and cut out (top); the model partially assembled and complete (left and middle); and a coloured version (right).

Net for Truncated Octahedron

14 faces (6 squares, 8 hexagons), 24 vertices, 36 edges. Octahedral (2, 3, 4-fold) symmetry.

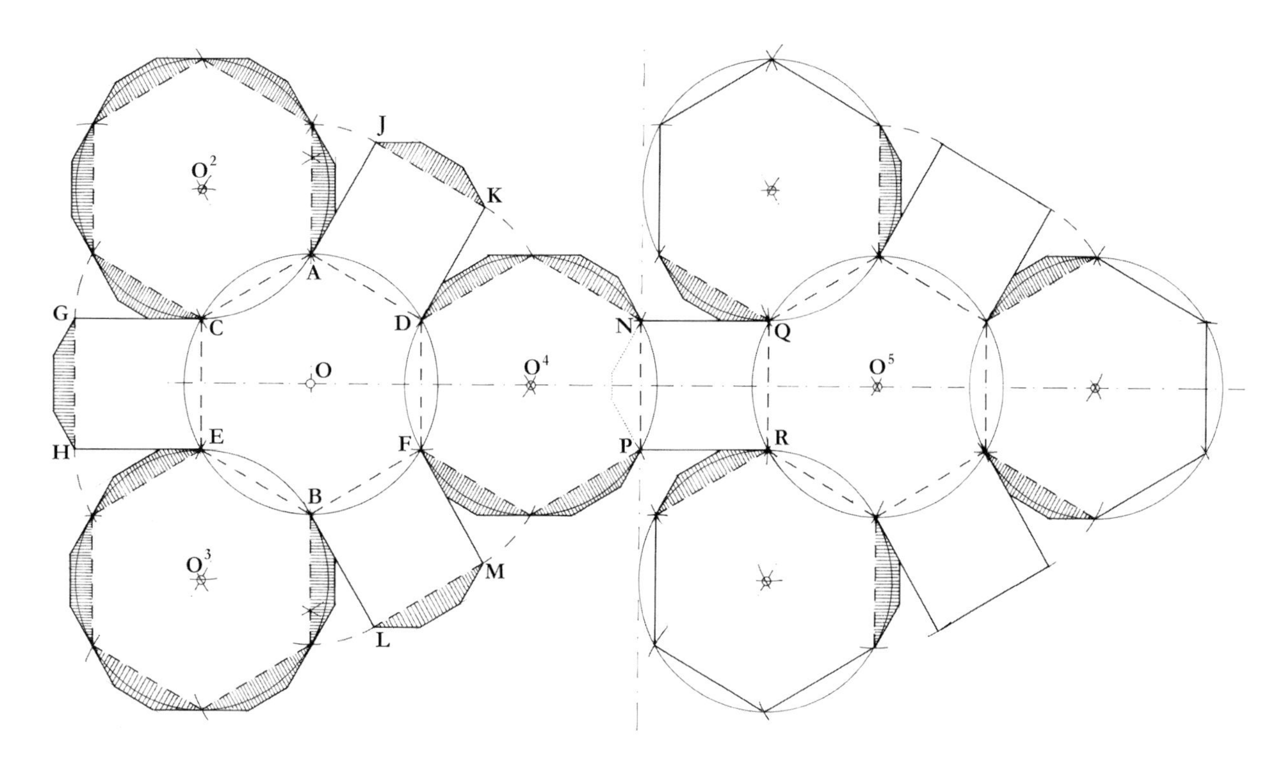

Fig. 5.7. Construction drawing.

Method:

1. On a long horizontal draw a circle centre O (the diameter of your circle should be just less than one sixth of the width of your sheet). Construct the vertical axis through O to give you points A and B. The rest of the construction is drawn using only the original radius.

2. Draw arcs from A to find C and D; and from B to find E and F. This gives you the first hexagon ADFBEC.

3. A line through DC extended to the left, and an arc from C will find G. Similarly, a line through FE and an arc from E will find H. GCEH is the first square. Find squares AJKD and FMLB in the same way by extending EA, BD; and AF, CB.

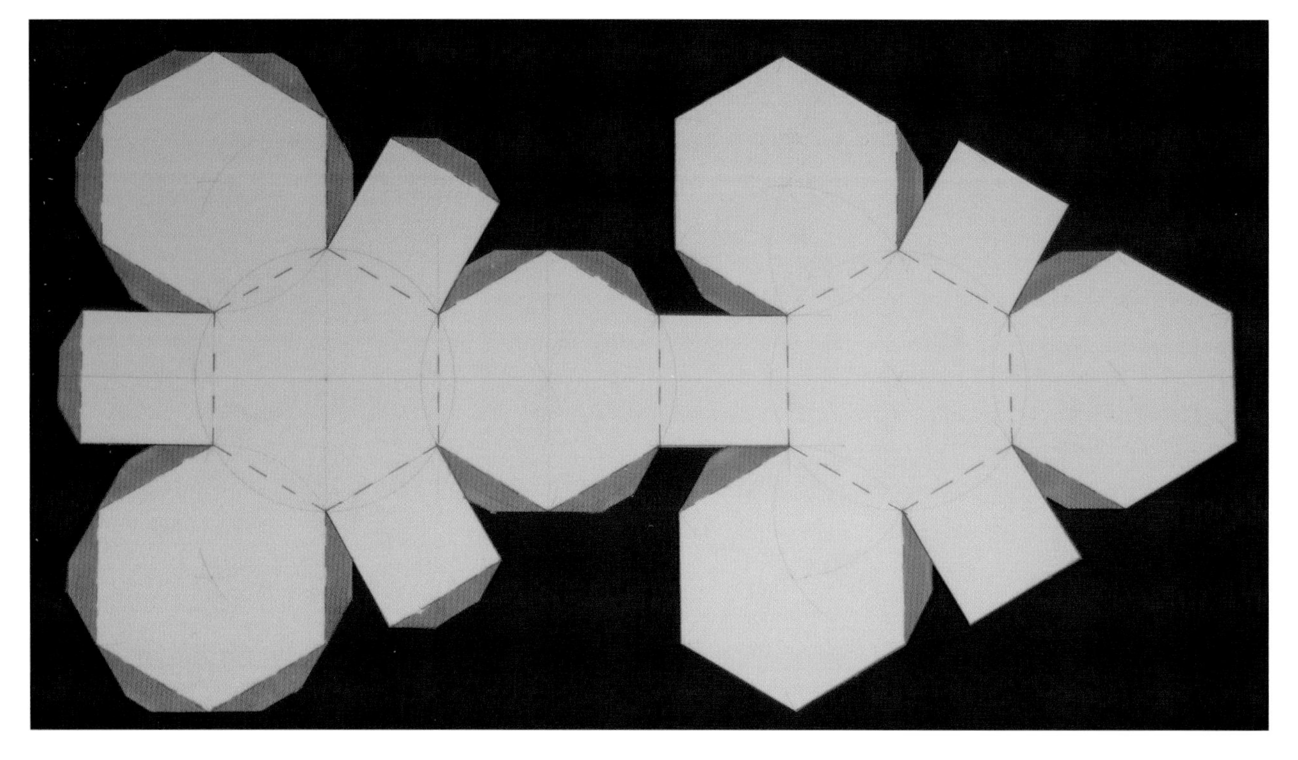

Fig. 5.8. Illustration of the net drawn out on card and cut out.

4. Pairs of arcs from A and C, B and E, D and F will cross to find new centres O^2, O^3 and O^4. On each of these new centres draw a circle, and walk the radius round each circle to find the points of the hexagons.

5. Square NQRP is found by extending lines through CD and EF to the right, and drawing arcs from N and P to find Q and R. Centre O^5 is found by drawing arcs from Q and R to cross. With this new centre established, the rest of the figure is constructed in the same way as the first half.

6. Add tabs and mark the cut and fold lines as shown.

7. Cut out along the solid lines: score along the dashed lines. Fold upwards along the scored lines.

8. Glue carefully and progressively (tabs inside) – one or two tabs at a time. Clean up and decorate as you wish.

Note:

The net can be drawn in two halves, both identical apart from the tabs. You will also need one additional tab to join the two halves together. The dotted line on Figure 5.7 between points N and P shows this.

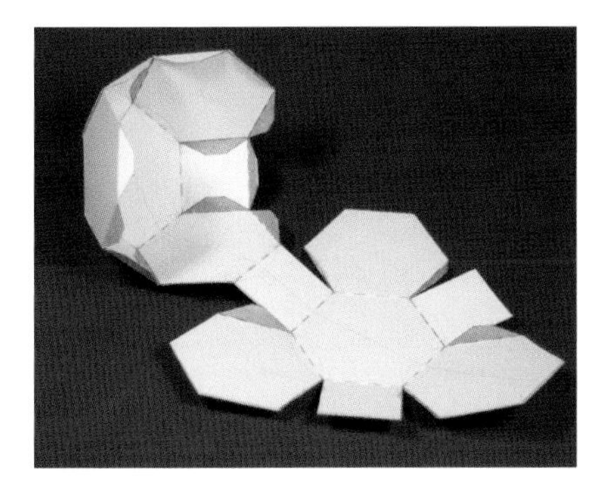

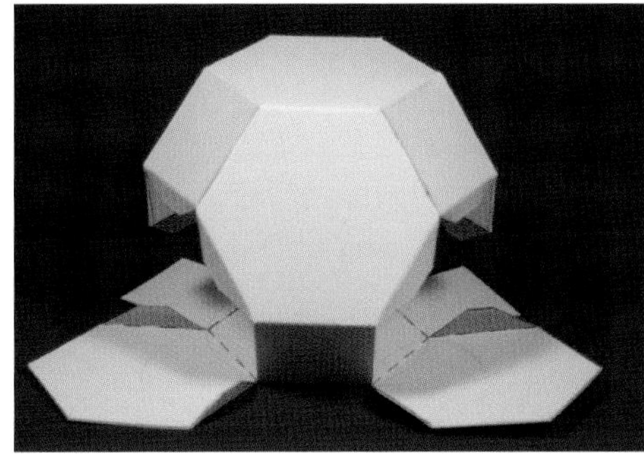

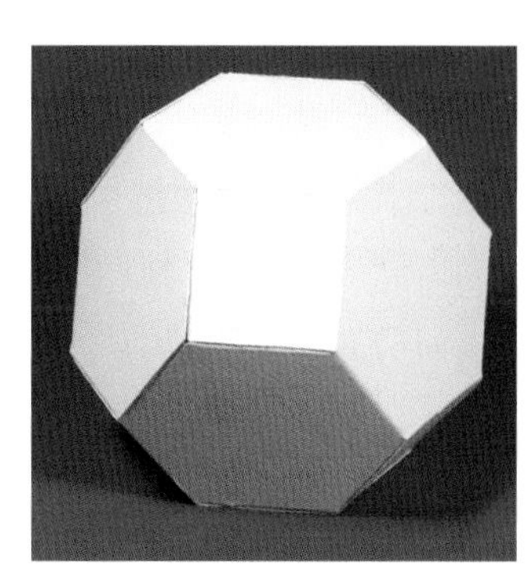

Figs. 5.9–5.12. Illustrations of the model partially assembled (above left and middle); complete (above right); and a coloured version (below right).

Net for the Cuboctahedron (Dymaxion)

14 faces (8 triangles, 6 squares), 12 vertices, 24 edges. Octahedral (2, 3, 4-fold) symmetry.

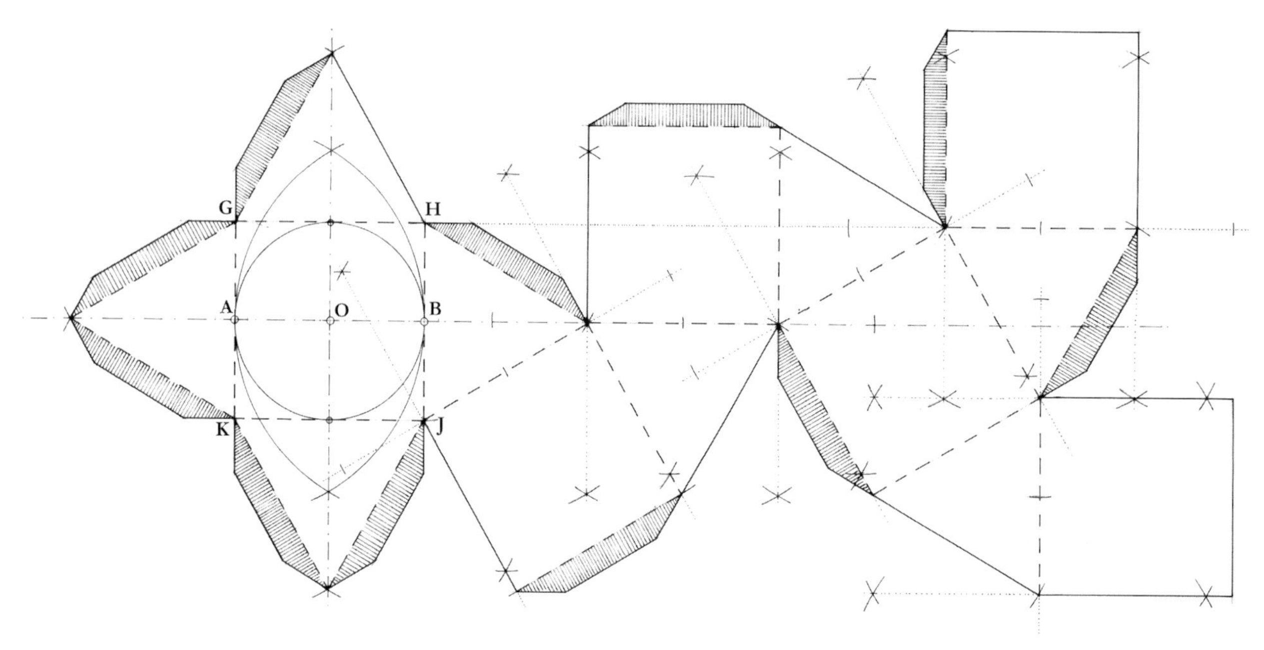

Fig. 5.13. Construction drawing.

Method:

1. Follow the steps below to draw first the initial square GHJK (Fig. 5.14), and then adjoining equilateral triangles (see Figure 5.15) and squares (see Figure 5.16) to create the whole construction (Fig. 5.13 above).

2. Start with AB on a horizontal line (Fig. 5.14). AB is the side length of the solid. Draw arcs from A and then B with radius AB to intersect above and below. Join points C and D of the vesica and extend to give the vertical axis through O (the midpoint of AB). Draw a circle centred on O radius OA, intersecting the vertical axis at E and F. From A, E, B and F in turn draw arcs radius OA to cross at the four corner points of the square GHJK.

3. Equilateral triangles can be constructed on a side by drawing arcs in turn from each endpoint of the line (see Figure 5.15).

4. Squares can be constructed on a side (see Figure 5.16). Extend the line through AB on either side. With radius OA from the original square draw circles on A and B. These will give you points D, C (midpoint of AB) and E. With radius DC (=AB) draw arcs from D, C and E to cross above and below. This will give you points F and H perpendicular to A, and G and J perpendicular to B. Draw AF and extend beyond F: and BG and extend beyond G. With radius AB, arc from A to cut AF extended at K, and from B to cut BG extended at L. AKLB is our square.

5. Add tabs and mark the cut and fold lines as shown.

6. Cut out along the solid lines: score along the dashed lines. Fold upwards along the scored lines. Glue carefully and progressively (tabs inside) – one or two tabs at a time. Clean up and decorate as you wish.

Note:

The Cuboctahedron was called the Dymaxion by Buckminster Fuller.

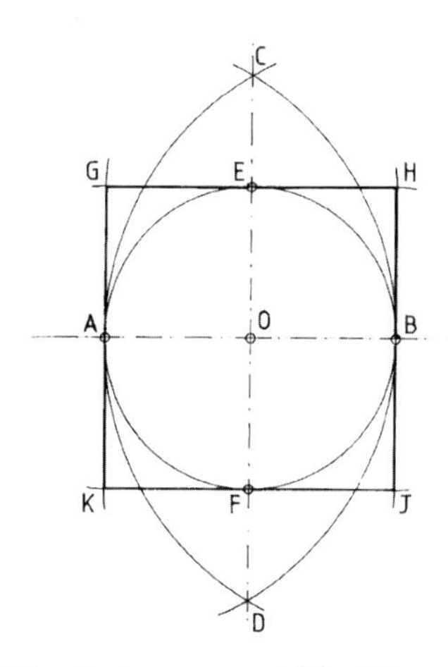

Fig. 5.14. The first square, width AB.

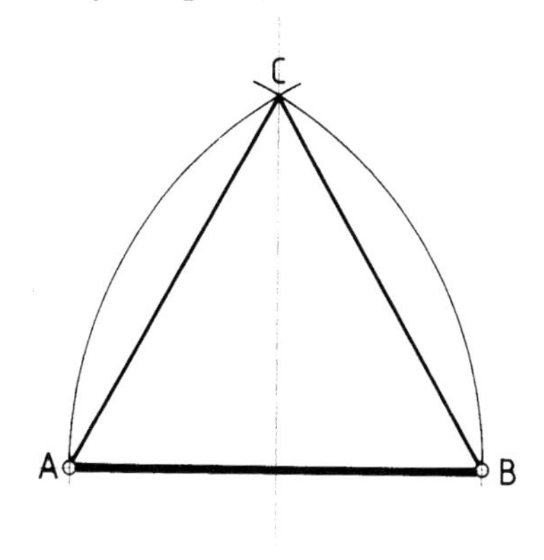

Fig. 5.15. An equilateral triangle on AB.

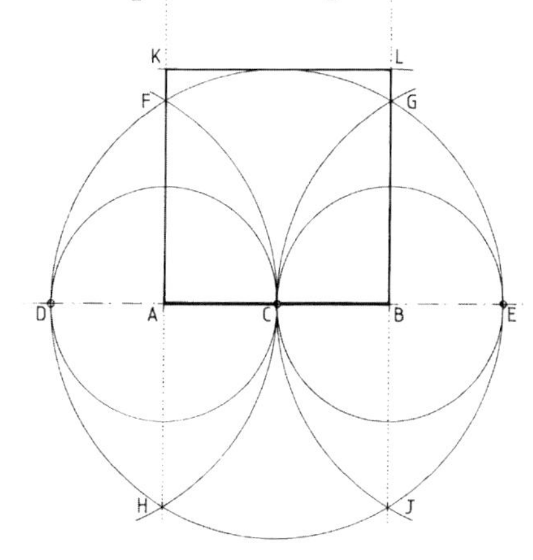

Fig. 5.16. A square on AB.

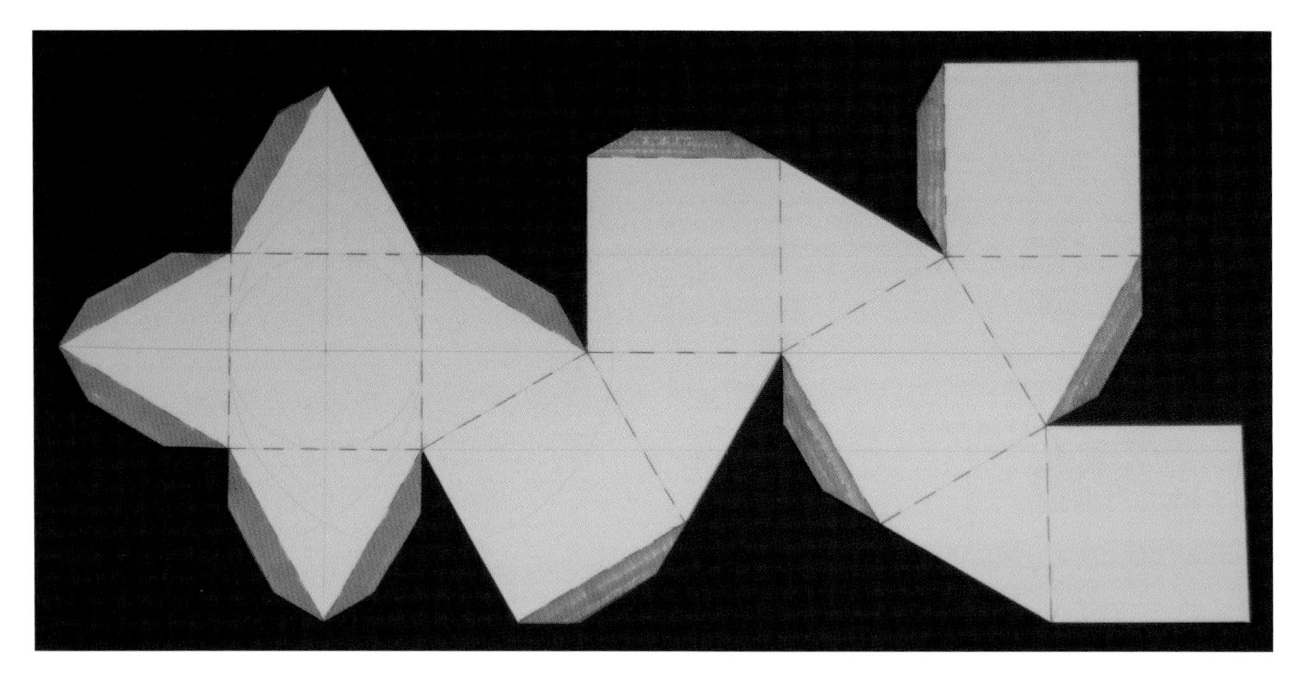

Fig. 5.17. Illustration of the net drawn out on card and cut out.

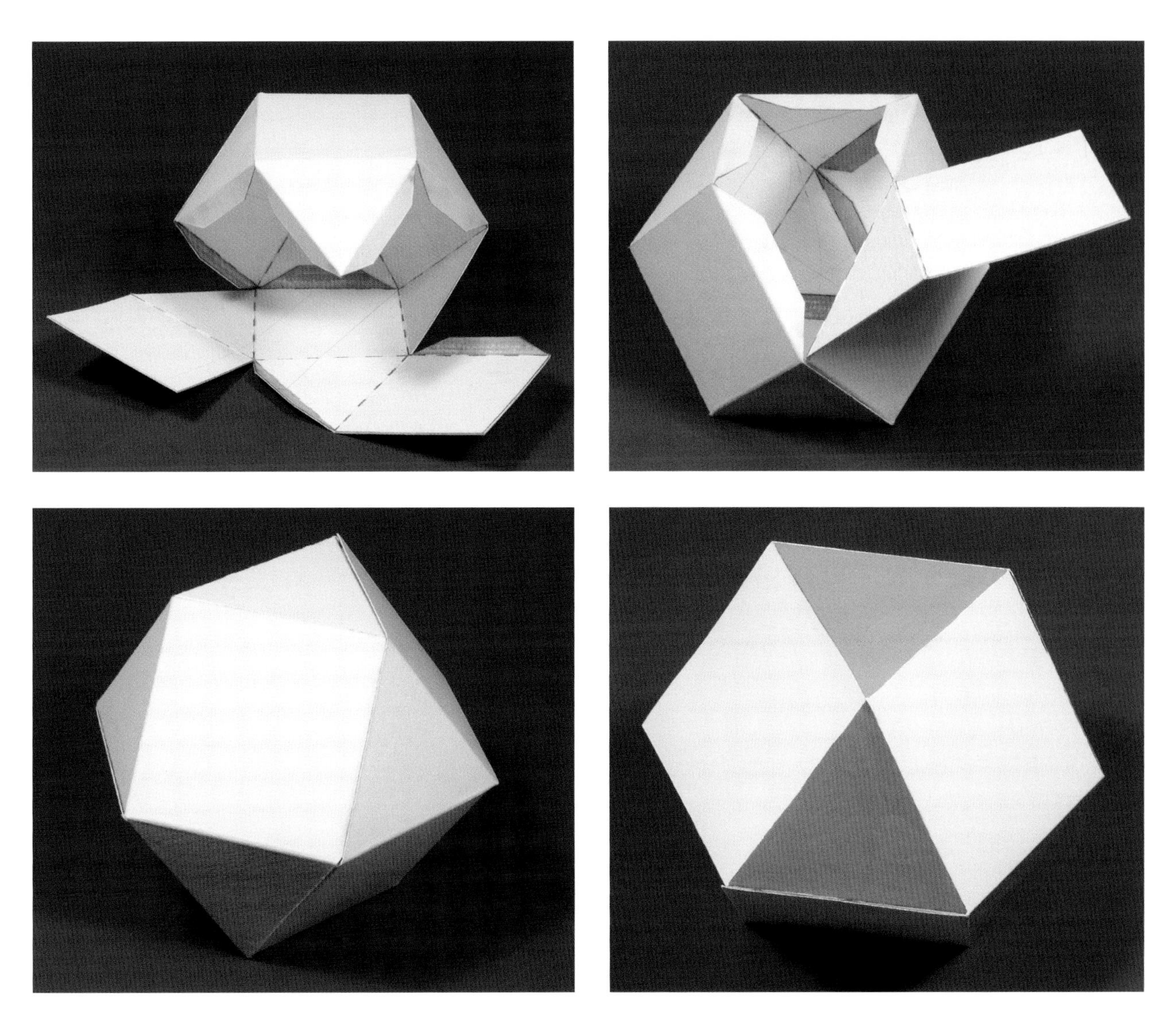

Figs. 5.18–5.21. Illustrations of the model partially assembled (top); complete (bottom left); and coloured (bottom right).

Net for Truncated Cube

14 faces (8 triangles, 6 octagons), 24 vertices, 36 edges. Octahedral (2, 3, 4-fold) symmetry.

Method:

1. (Step 1) Draw four circles on a horizontal. Construct verticals through each circle centre and through each diameter point on the edges of the circles. Draw tangents to the four circles above and below to cross the verticals, forming four touching squares (Fig. 5.22).

2. (Step 2) To complete the first octagon (left-most circle in Figure 5.23 below) draw a circle through the corners of the surrounding square, and join the four points where the new larger circle crosses the horizontal and vertical axes. Where this (dynamic) square – which is the same size as the first (static) square – overlaps the first square, an octagon is formed.

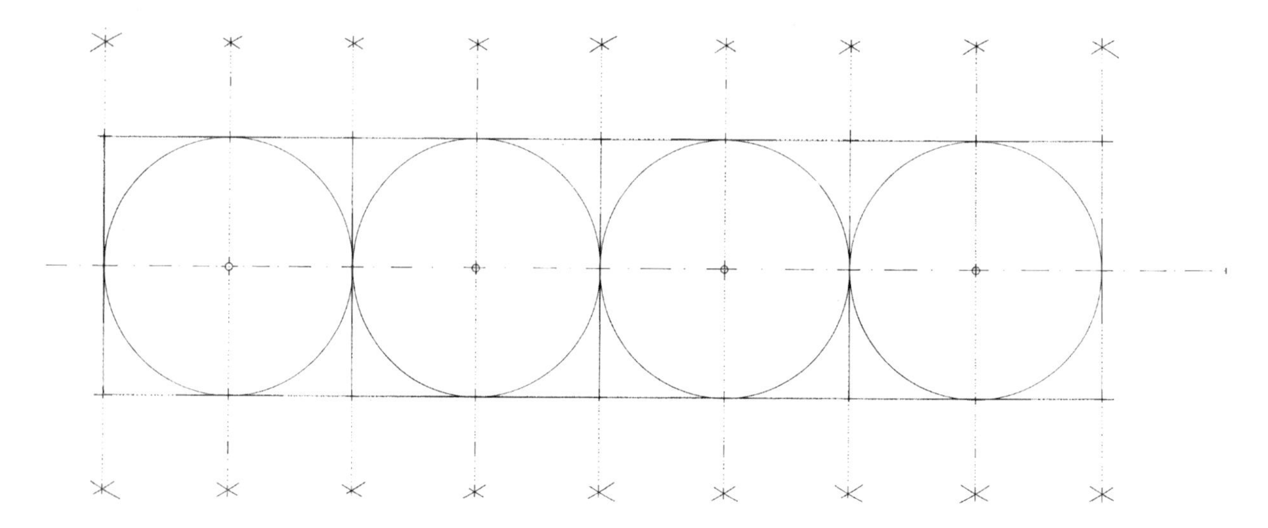

Fig. 5.22. Step 1.

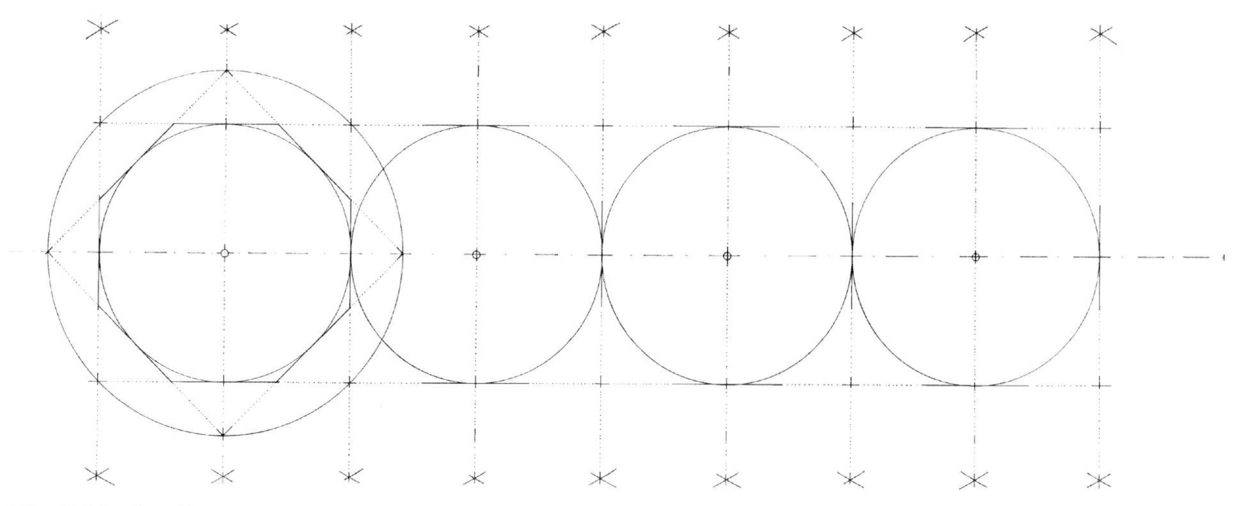

Fig. 5.23. Step 2.

3. (Step 3) Repeat to form octagons around the other three circles. Construct equilateral triangles on the sides shown (see Figure 5.15, page 49).

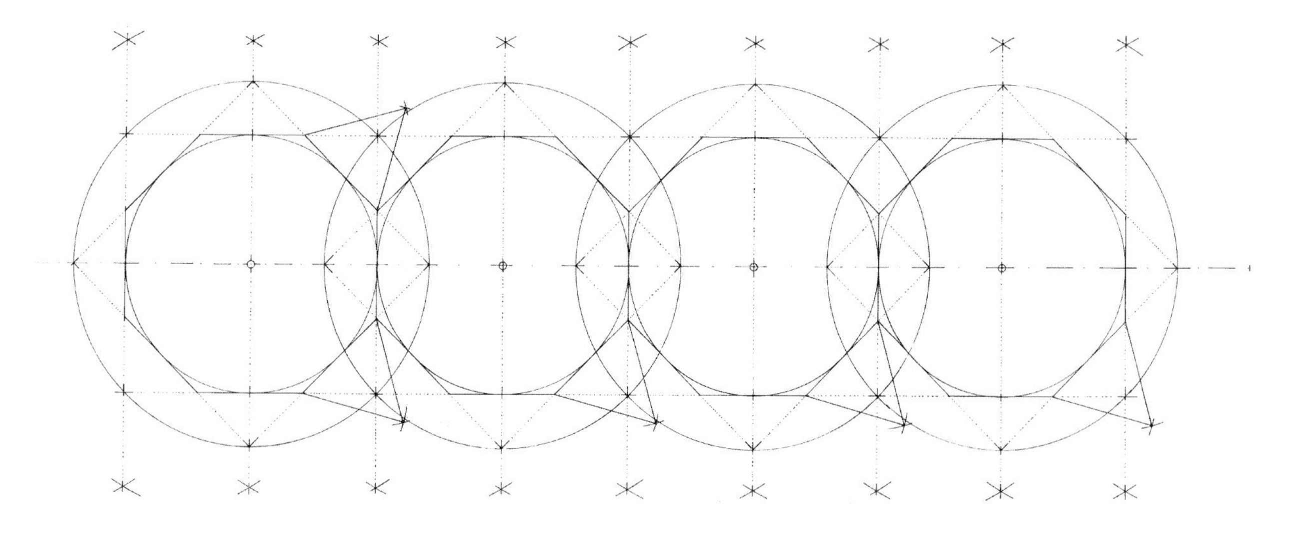

Fig. 5.24. Step 3.

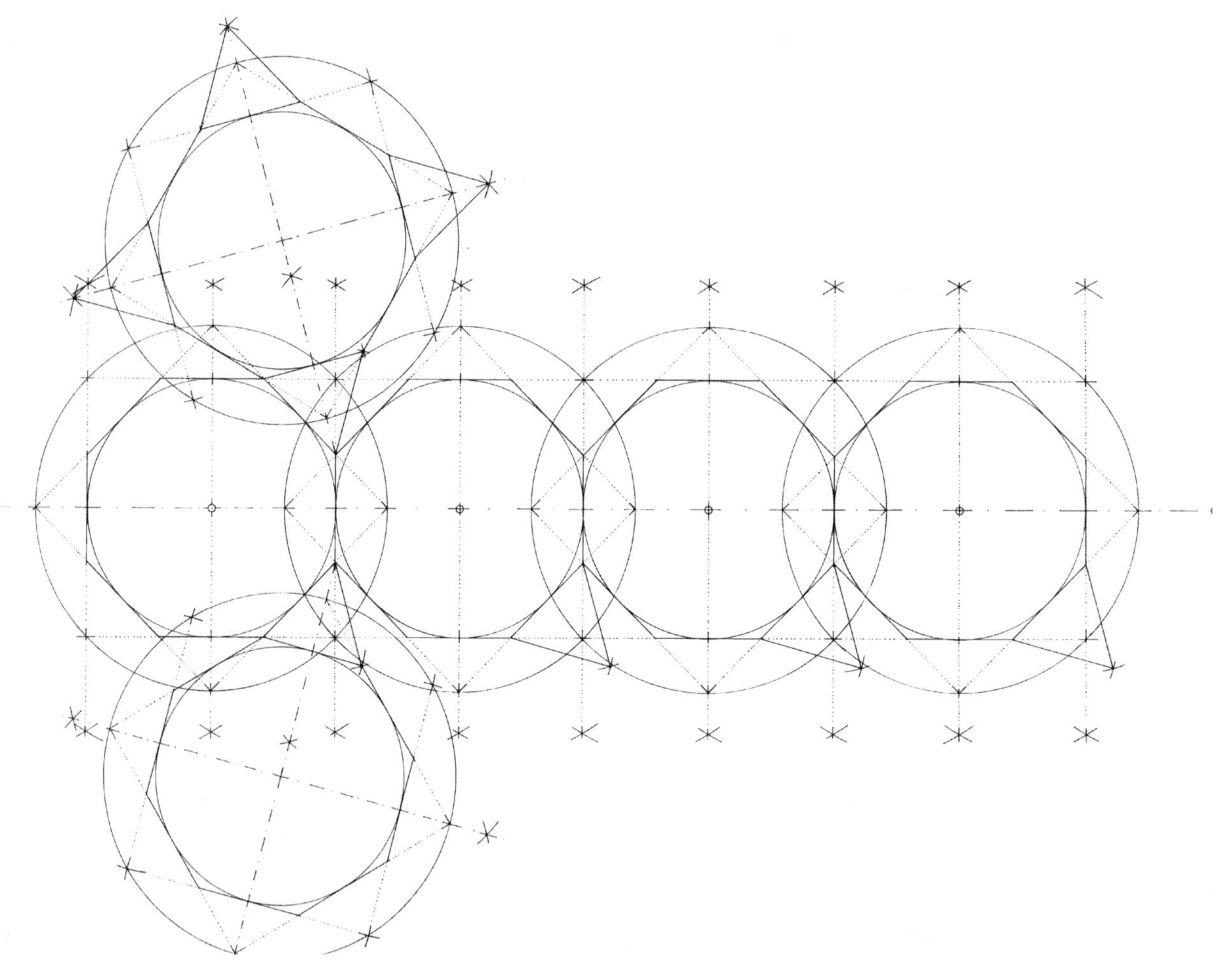

Fig. 5.25. Step 4.

4. (Step 4) Construct an octagon on the side of each of the two triangles shown. Add the final three equilateral triangles to the top octagon.

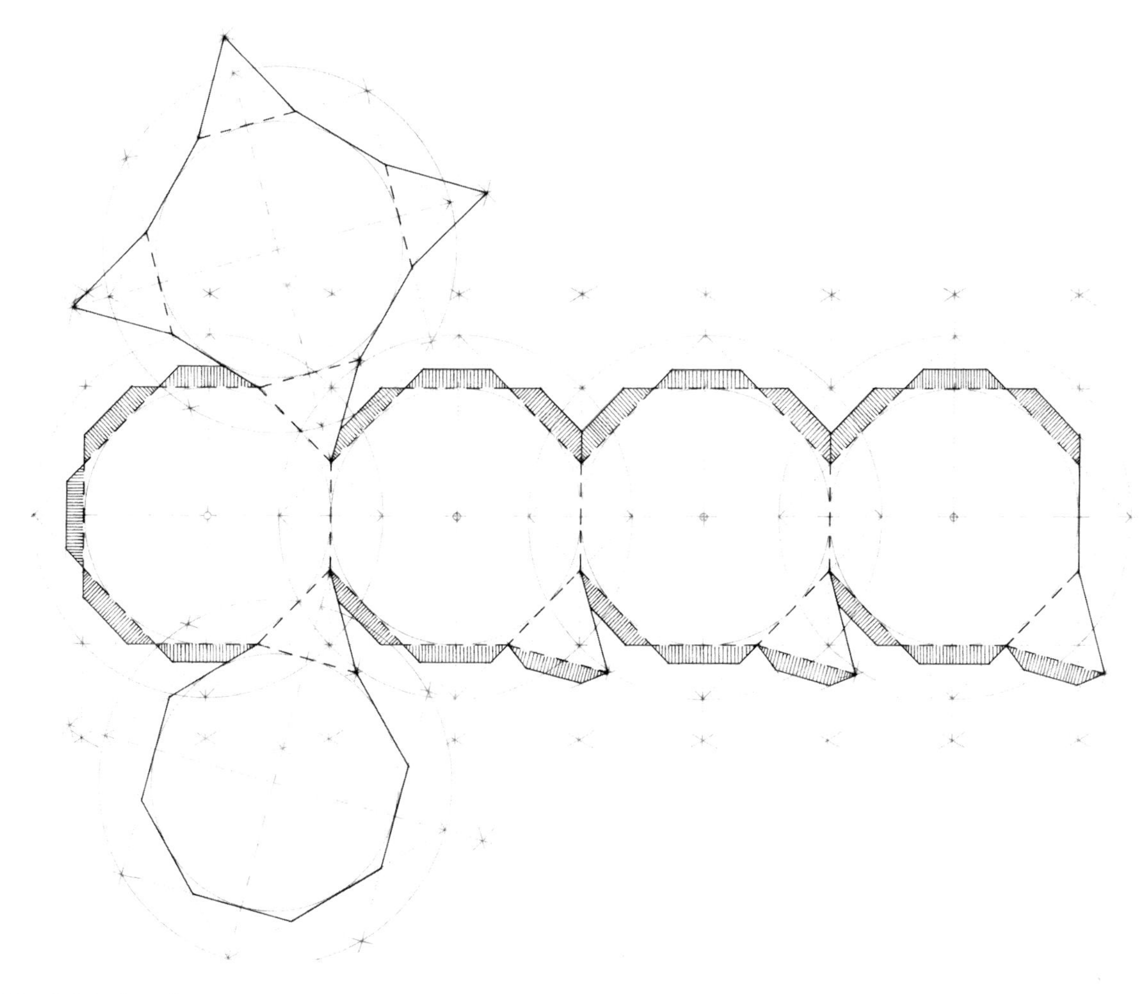

Fig. 5.26. Step 5.

5. (Step 5) Add tabs and mark the cut and fold lines as shown.

6. Cut out along the solid lines: score along the dashed lines. Fold upwards along the scored lines. Glue carefully and progressively (tabs inside) – one or two tabs at a time. Clean up and decorate.

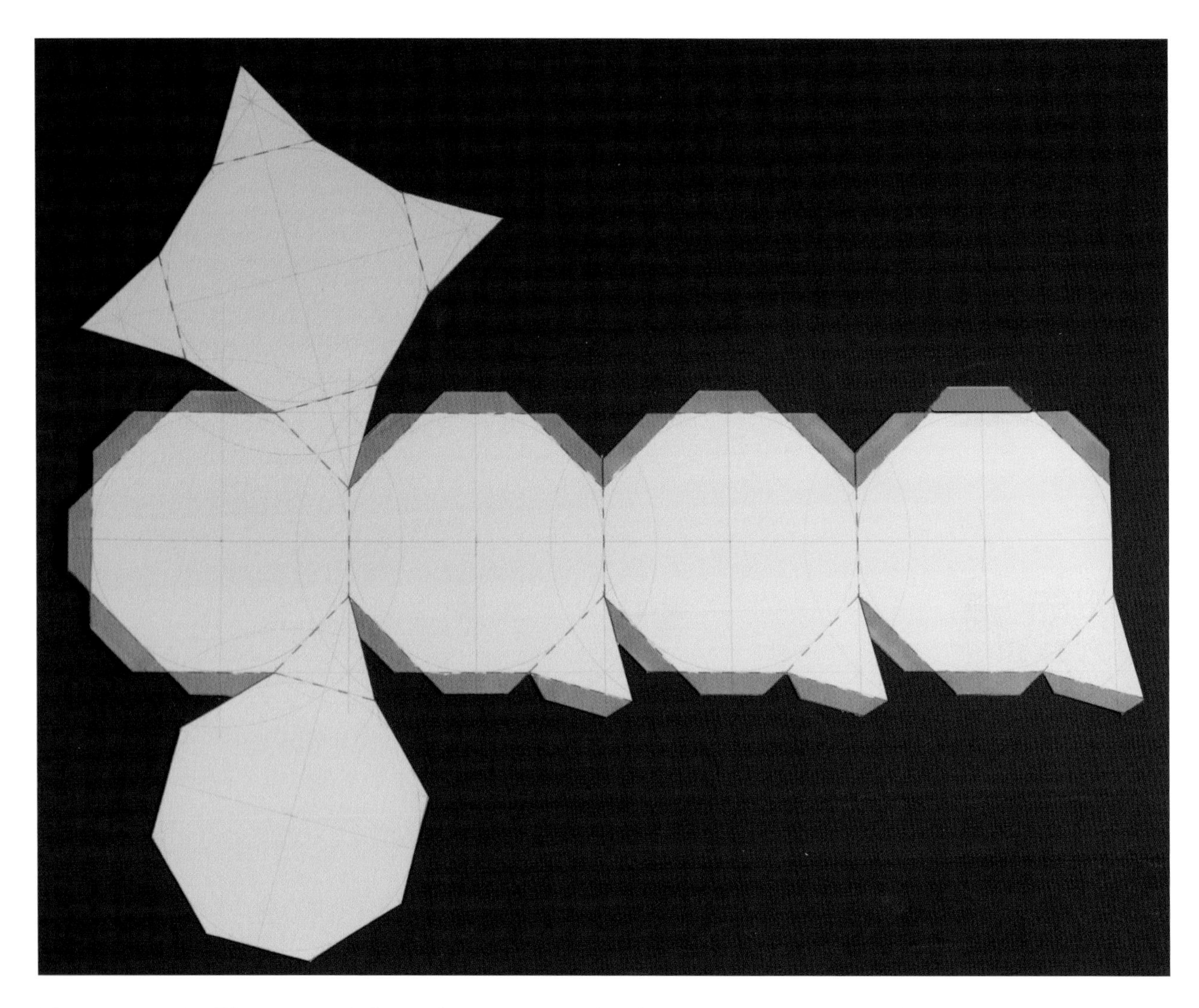

Figs. 5.27–5.28. Illustrations of the net drawn out on card and cut out (above) and the model partially assembled (facing page top left).

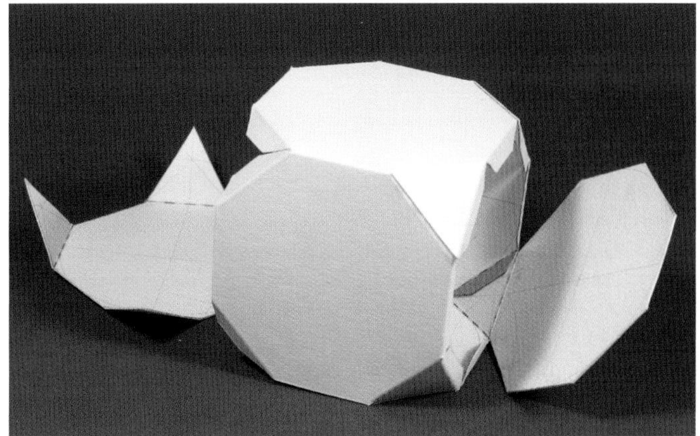

Fig. 5.29. The model further assembled.

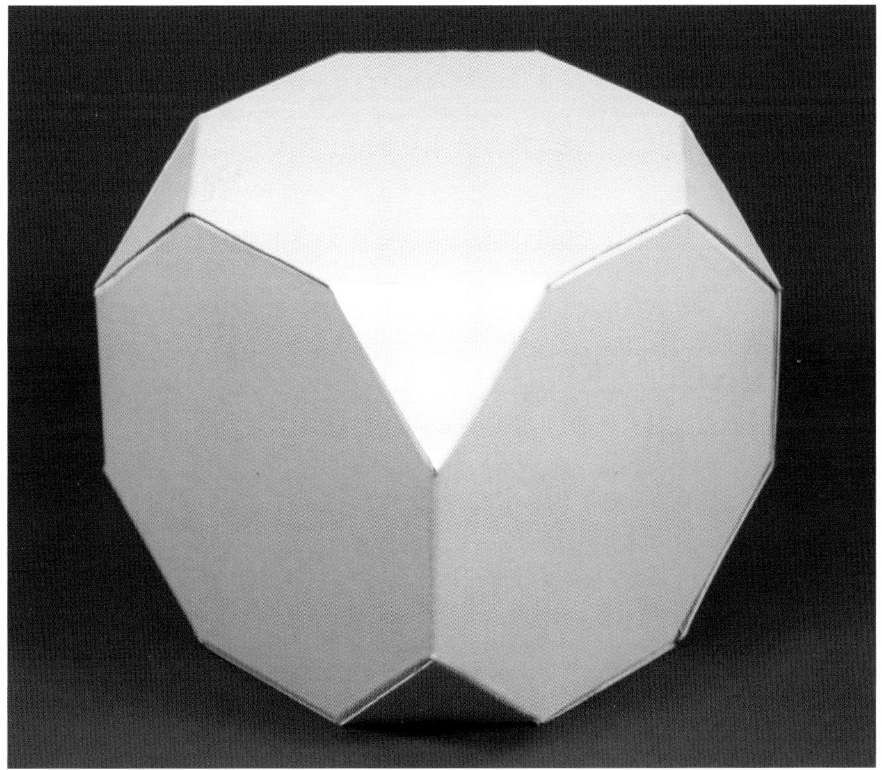

Fig. 5.30. The model complete.

Net for Rhombicuboctahedron

26 faces (8 triangles, 18 squares), 24 vertices, 48 edges. Octahedral (2, 3, 4-fold) symmetry.

Method:

1. (Step 1) Mark eight measures along a horizontal line drawn just below the middle of your sheet. Construct perpendiculars at each mark (Fig. 5.31).

2. (Step 2) With the same measure, mark off on each vertical two lengths above and one below the horizontal, with one extra above and below in column 5. Join up the marks to produce an 8 by 3 grid, plus the two extra squares, as shown (Fig. 5.32).

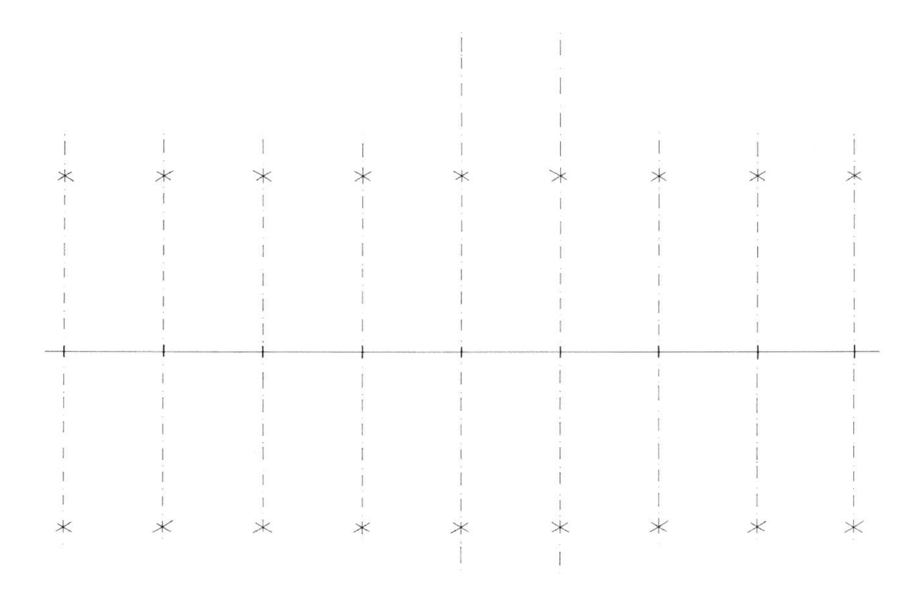

Fig. 5.31. Step 1.

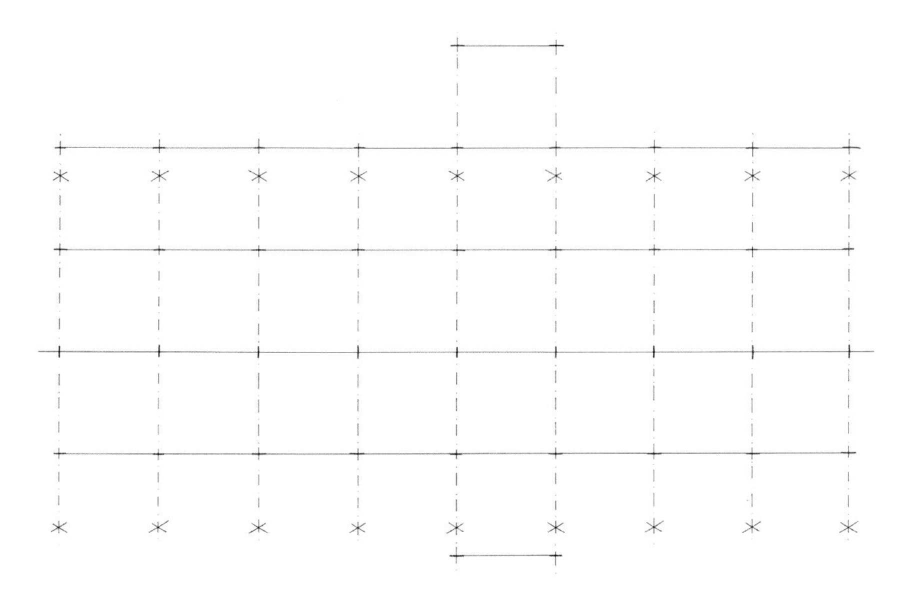

Fig. 5.32. Step 2.

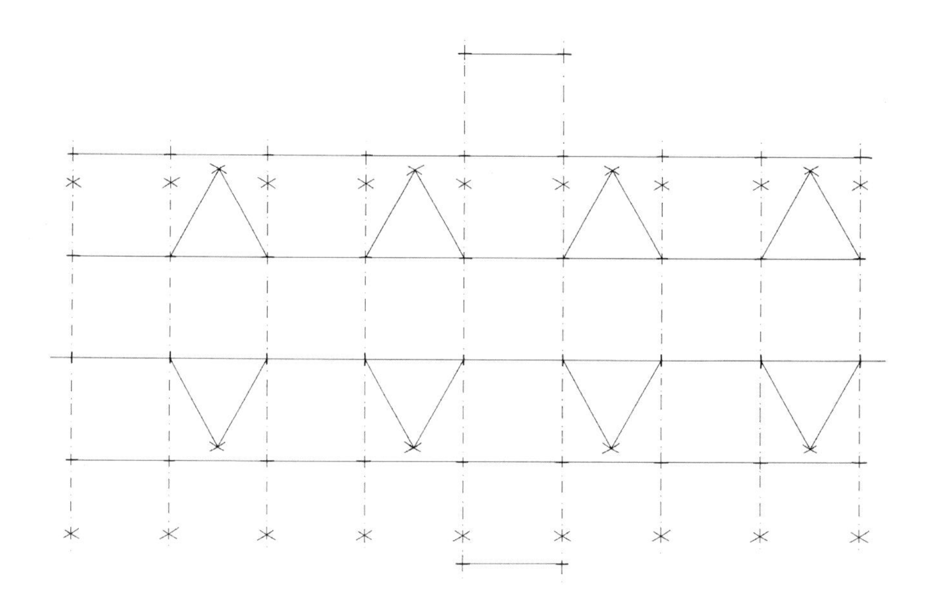

Fig. 5.33. Step 3.

3. (Step 3) Construct equilateral triangles in the eight positions shown (Fig. 5.33).

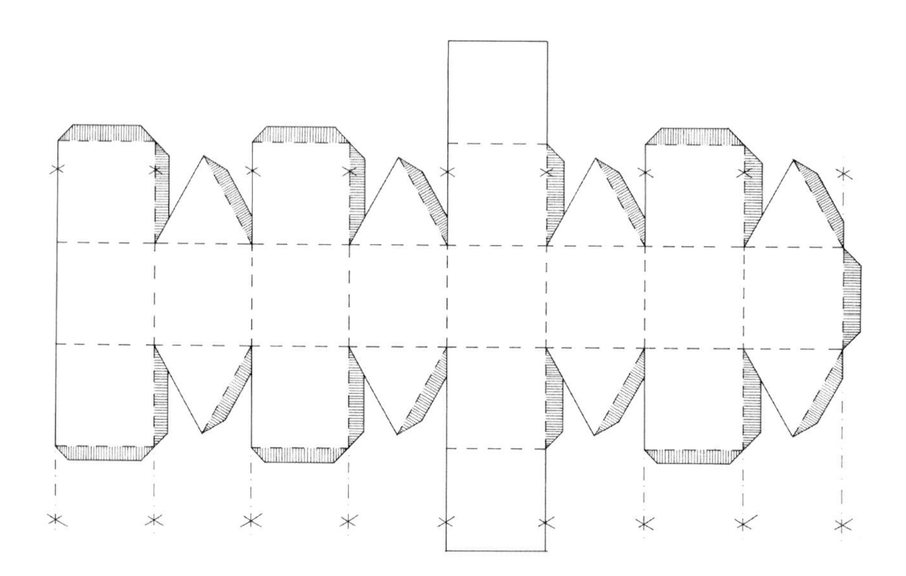

Fig. 5.34. Step 4.

4. (Step 4) Add tabs and mark the cut and fold lines as shown, to complete the net (Fig. 5.34).

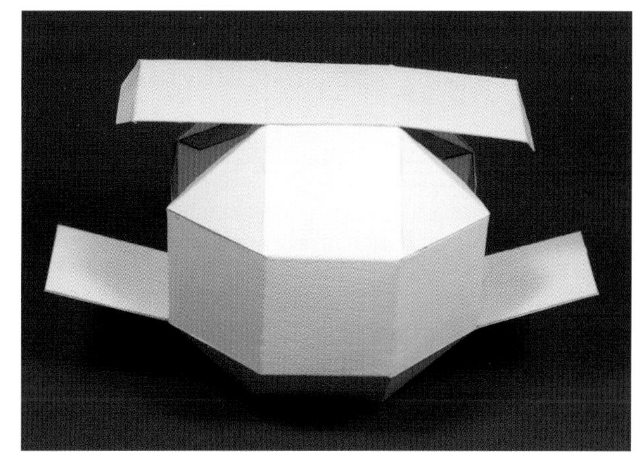

Figs. 5.35–5.37 Illustrations of the model partially assembled (above left and right); and complete (below).

5. Cut out along the solid lines: score along the dashed lines. Fold upwards along the scored lines. Glue carefully and progressively (tabs inside) – one or two tabs at a time. Clean up and decorate as you wish.

Net for Truncated Cuboctahedron

26 faces (12 squares, 8 hexagons 6 octagons), 48 vertices, 72 edges. Octahedral (2, 3, 4-fold) symmetry.

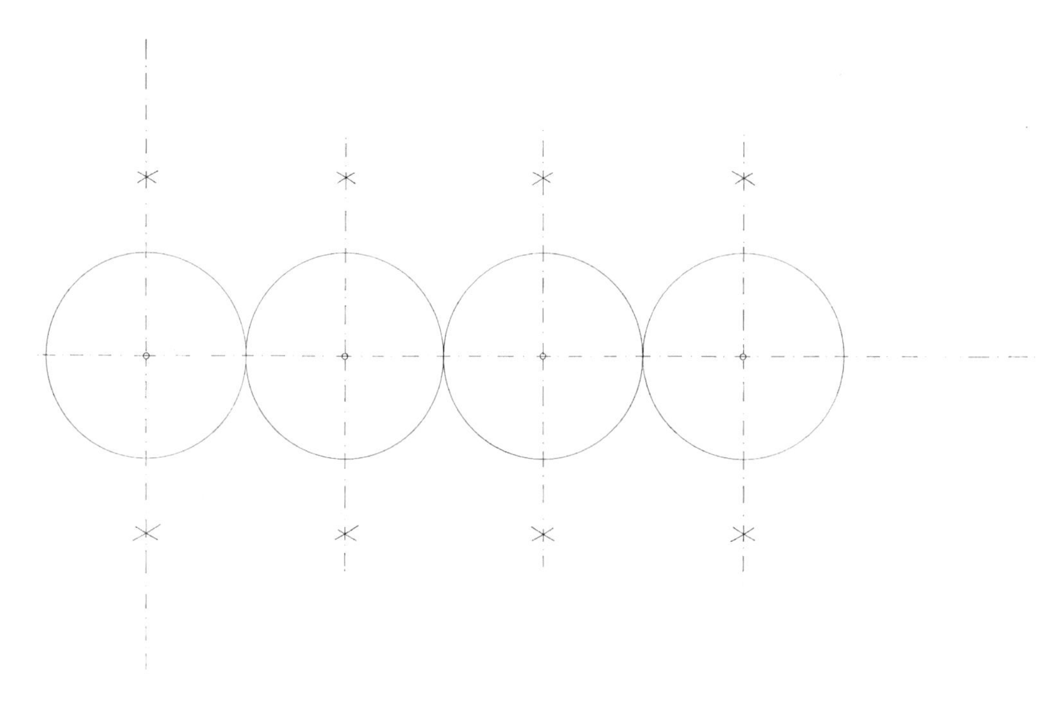

Fig. 5.38. Step 1.

Method:

1. (Step 1) On a horizontal draw four touching circles (with a bit of room – one radius length – to the right). Construct verticals through the four centre points by drawing arcs from the diameter points on the horizontal (Fig. 5.38).

2. (Step 2) On all four diameter points of each circle draw further circles with the original radius, and additional circles above and below on the first vertical as shown (Fig. 5.39).

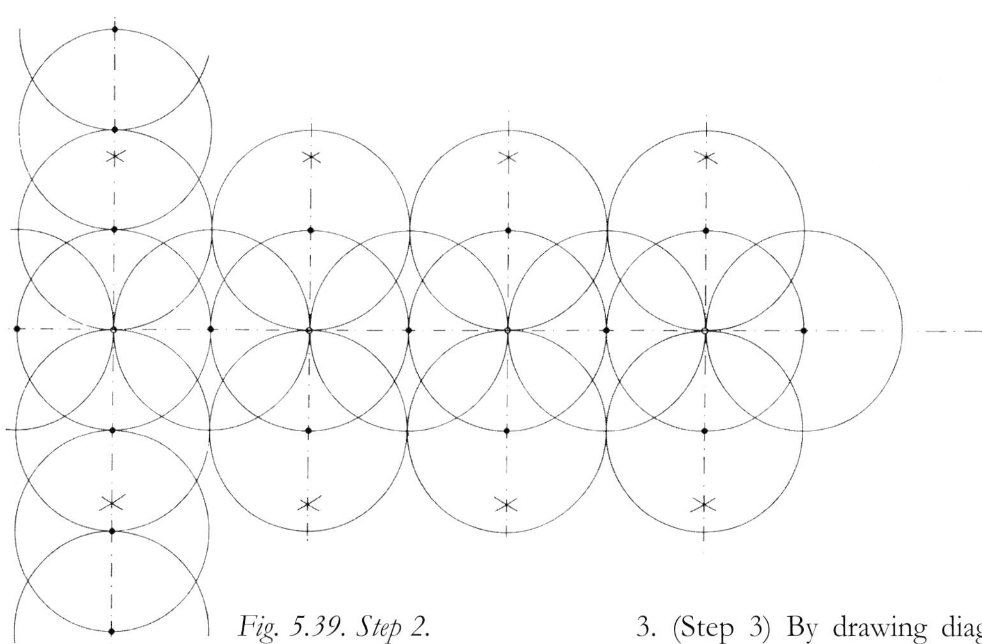

Fig. 5.39. Step 2.

3. (Step 3) By drawing diagonal lines between diameter points (circle centres), the diagonals of the required octagons are found (Fig. 5.40).

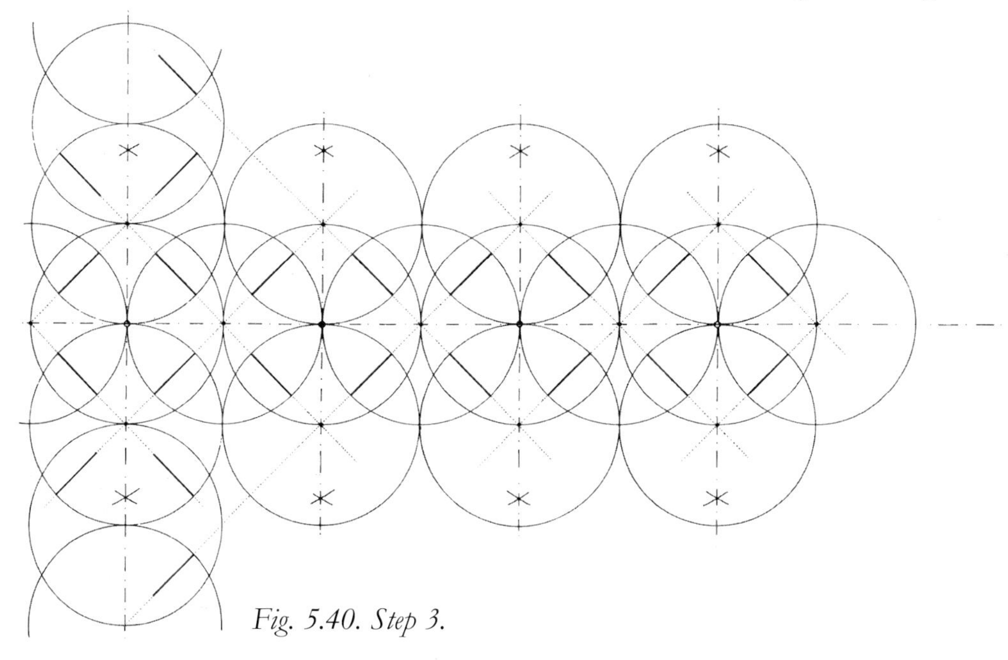

Fig. 5.40. Step 3.

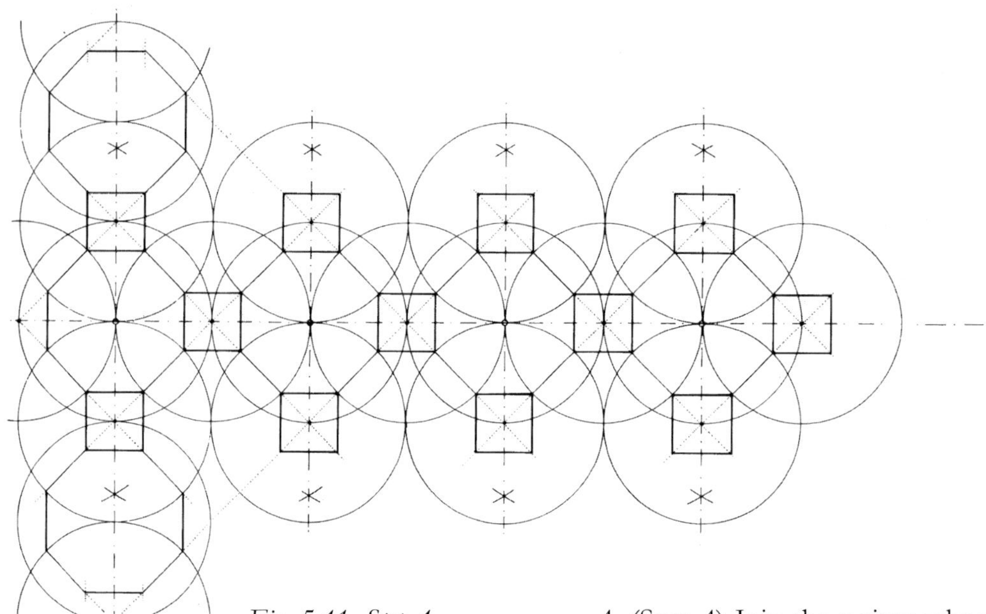

Fig. 5.41. Step 4.

4. (Step 4) Join the points where these diagonals meet the circles to form squares – which also complete the octagons (Fig. 5.41).

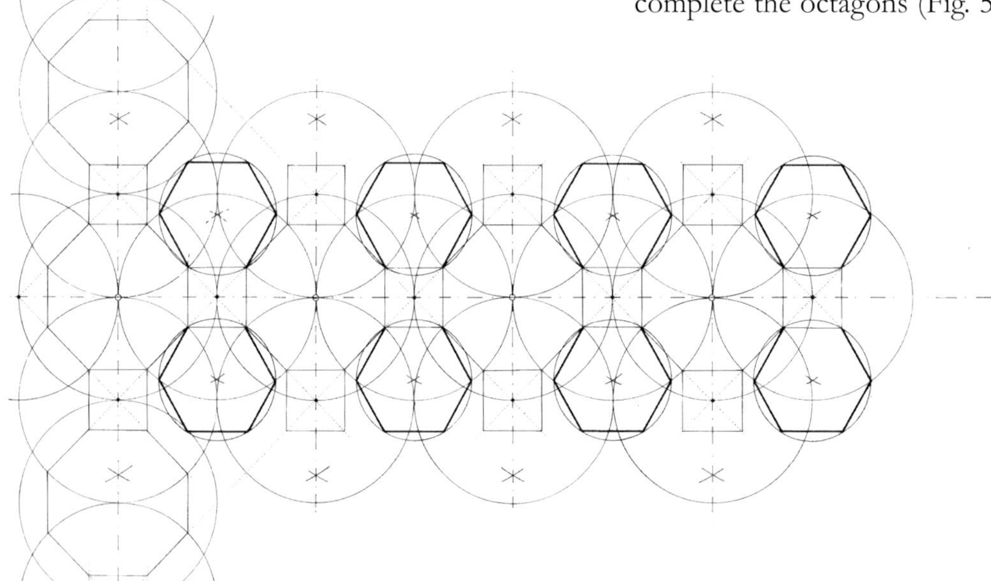

Fig. 5.42. Step 5.

5. (Step 5) Construct hexagons on the squares where shown (Fig. 5.42). The method is shown enlarged in the adjacent diagram (Fig. 5.43).

6. (Step 6) Add tabs and mark the cut and fold lines as shown (Fig. 5.44).

7. Cut out along the solid lines: score along the dashed lines. Fold upwards along the scored lines. Glue carefully and progressively (tabs inside) – one or two tabs at a time. Clean up and decorate as you wish.

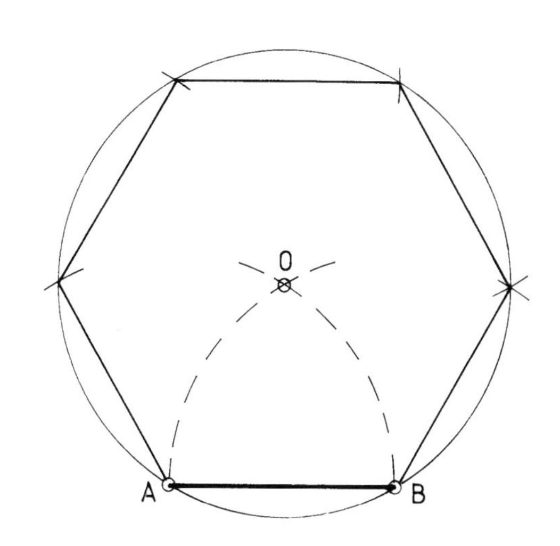

Fig. 5.43. Hexagon on a line.

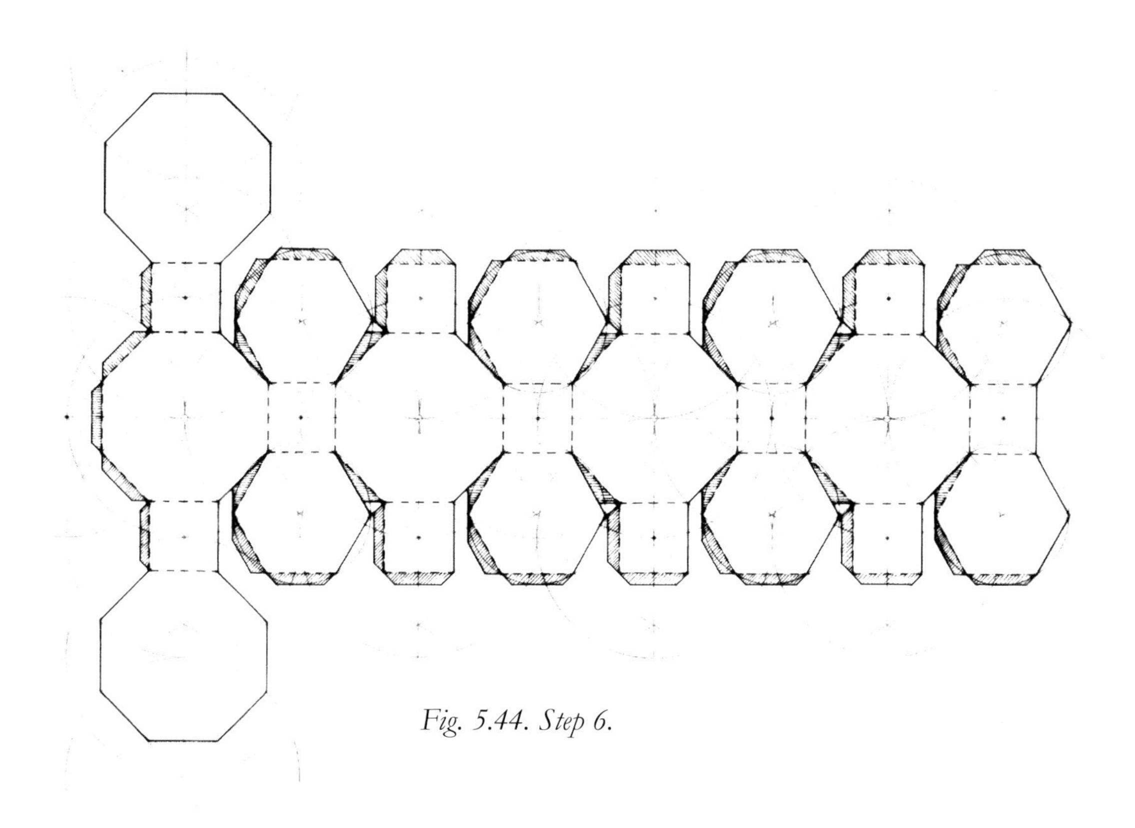

Fig. 5.44. Step 6.

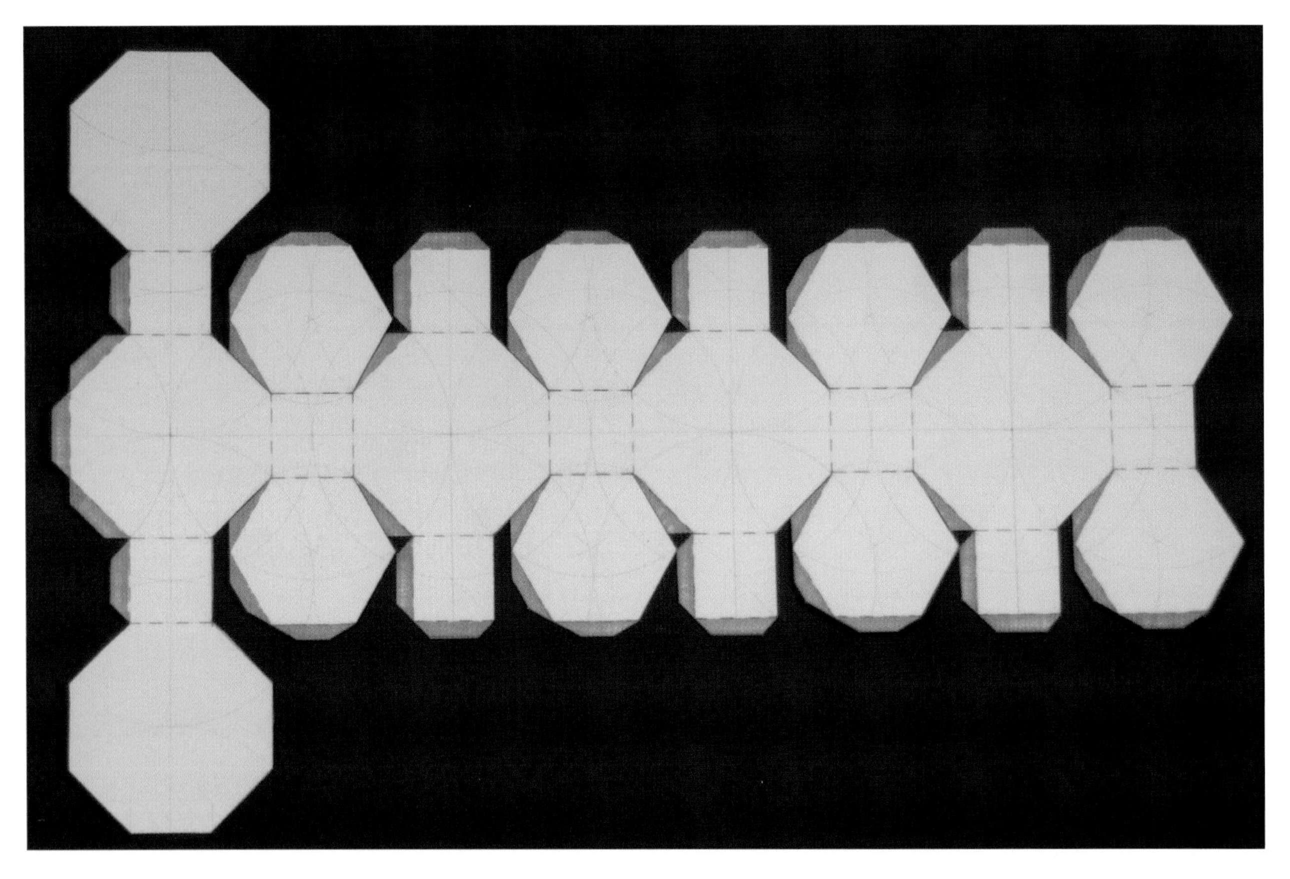

Figs. 5.45–5.47. Illustrations of the net drawn out on card and cut out (above); the model partially assembled (below left); and complete (below right).

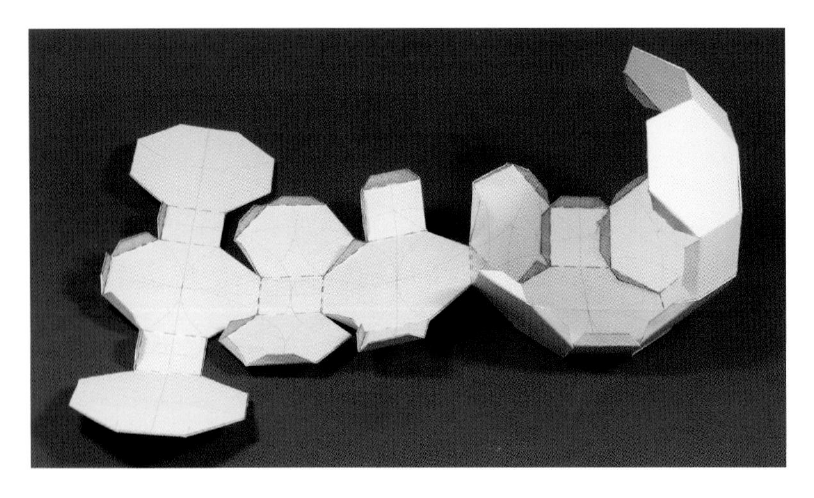

Net for Truncated Icosahedron

32 faces (12 pentagons, 20 hexagons), 60 vertices, 90 edges. Icosahedral (2, 3, 5-fold) symmetry.

Method:

This net is made in SIX parts:

1. PART 1. (Step 1) Having chosen your edge length AB on a horizontal line, construct the central pentagon as follows: draw two circles centred on A then B radius AB. Draw the vertical axis through the circle intersections C and D to find O (midpoint of AB). Draw a tangent across the top of the two circles, and from the point where this tangent crosses the vertical draw arcs either side radius OA. This gives upper corners E and F of square AEFB. With centre O and radius OE draw a semicircle to cross the horizontal at G and H. AB has now been extended on either side by a Golden Ratio amount. Now draw arcs from A and B with radius AH (=BG). Where they cross on the vertical axis is the upper point J of our pentagon. The other two points are given where the AH and BG arcs cross the first circles at L and K. Our pentagon is AKJLB.

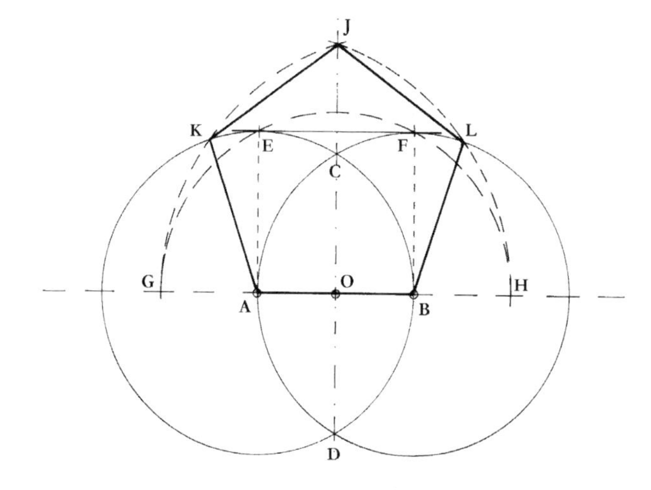

Fig. 5.48. Step 1.

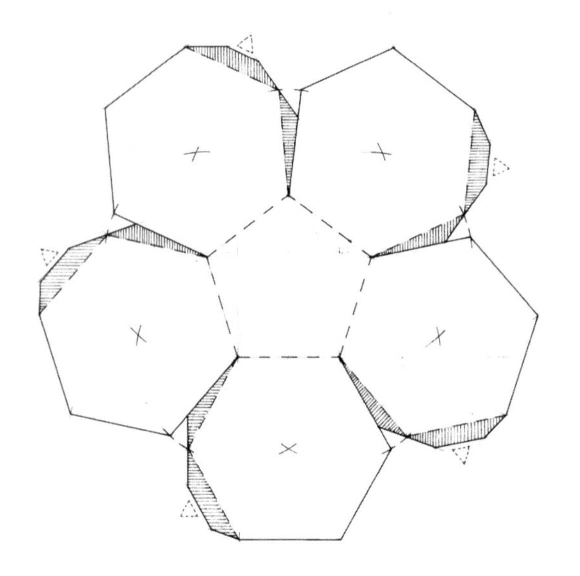

Fig. 5.49. Step 2.

2. (Step 2) On each side of the pentagon construct a hexagon – (see Figure 5.43, page 65). The arrows in the diagram indicate where further sections of the net will be attached.

3. PARTS 2-5 (Step 3) Starting with a hexagon (constructed as before) with the same edge length as used in the first drawing, add to the sides one hexagon and two pentagons as shown (Fig. 5.50). The pentagons can be more easily drawn this time (see Figure 5.51 above right) by using the previously drawn pentagon: measure with your compasses the length of the diagonal (AJ in Fig. 5.48) and draw an arc from each end of the base line. This gives you the apex of the pentagon. Now adjust your compasses to the edge length, and draw an arc from the end of the base line to cross arcs on either side from the apex to find the other two vertices. Finally, one further hexagon is added to the lefthand pentagon to complete this section of the net.
   You will need FOUR of these.

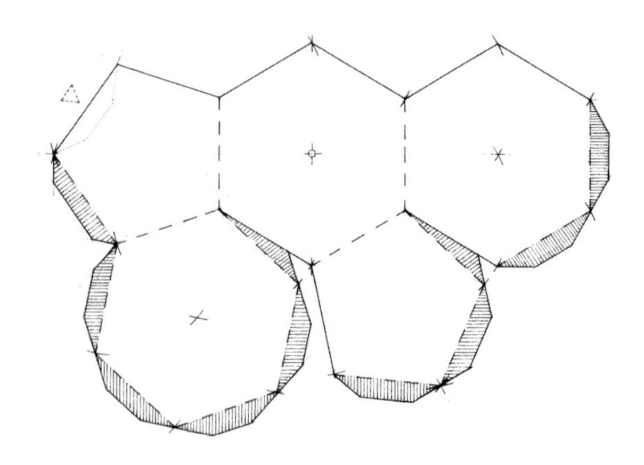

Fig. 5.50. Step 3.

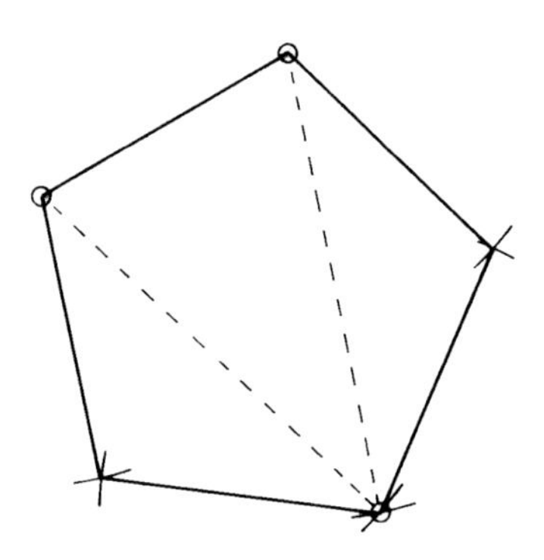

Fig. 5.51. Construction of pentagon using arcs of known measurements.

4. PART 6. (Step 4) is similar to the previous ones, but with an extra pentagon on one end (Fig. 5.52).

5. Add tabs, and cut and score lines, as shown.

6. Cut out along the solid lines: score along the dashed lines. Fold upwards along the scored lines. Glue carefully and progressively (tabs inside) – one or two tabs at a time. Clean up and decorate as you wish.

Fig. 5.53. Illustration of the six parts of the net drawn out on card and cut out.

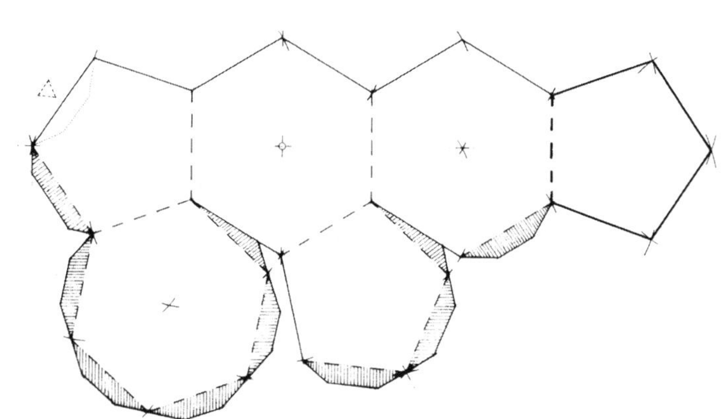

Fig. 5.52. Step 4.

Note:

Until 2006 the official World Cup football was made in the shape of a Truncated Icosahedron.

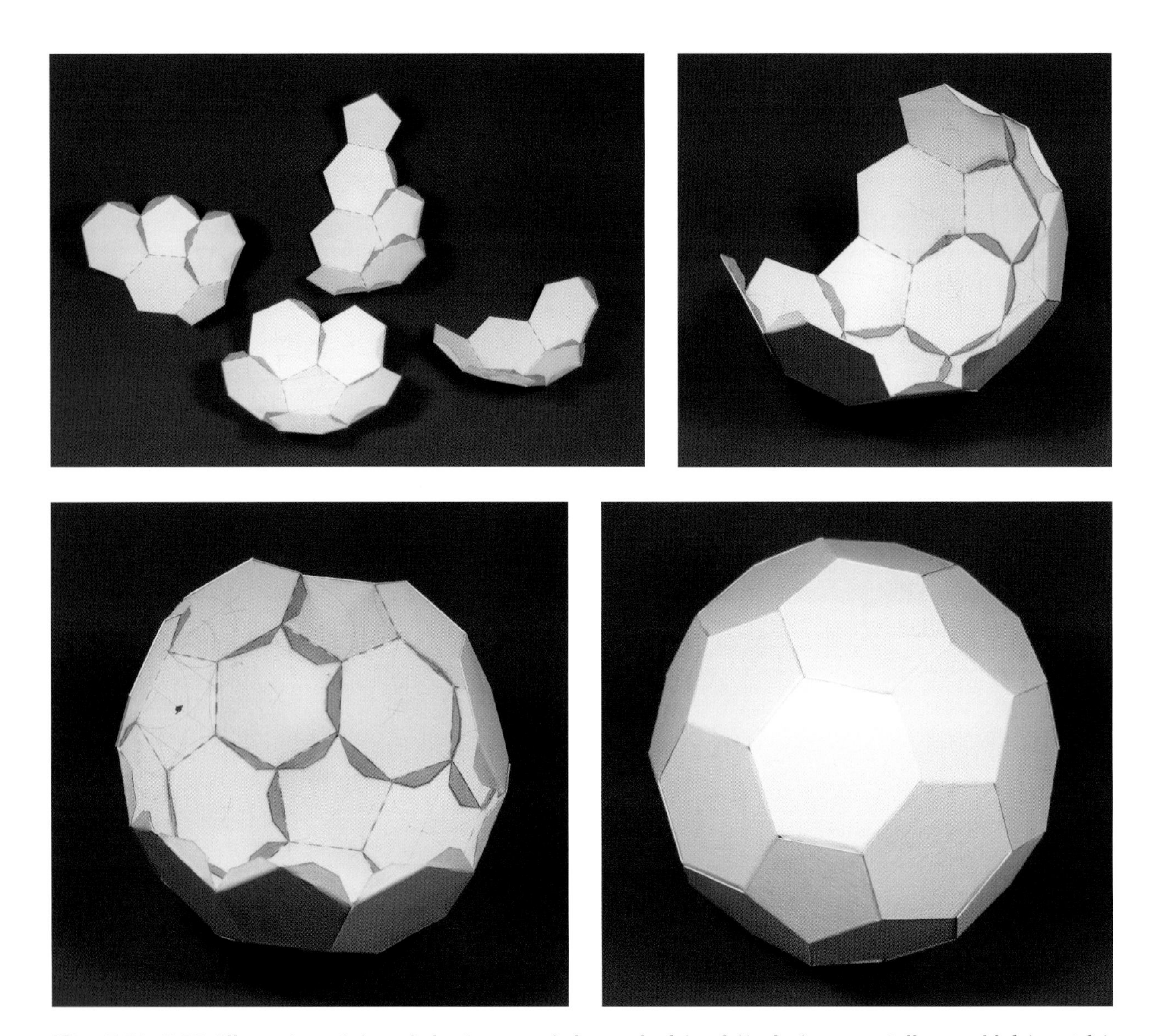

Figs. 5.54–5.57. Illustrations of four of the six parts of the net glued (top left); the figure partially assembled (top right); further assembled (bottom left); and complete (bottom right).

Net for Icosidodecahedron

32 faces (20 triangles, 12 pentagons), 30 vertices, 60 edges. Icosahedral (2, 3, 5-fold) symmetry.

Method:

This net is made in TWO parts.

1. PART 1. (Step 1) Start with a central pentagon (see page 67). Add an equilateral triangle to each side (see Figure 5.15 page 49). On one side of each triangle add a pentagon (see Figure 5.51 page 68) – the setting out arcs only are shown in Figure 5.58.

2. (Step 2) Add further equilateral triangles to the appropriate sides of the pentagons: observe the direction of 'spin' (Fig. 5.59).

3. (Step 3) Complete the first part of the net by adding tabs and marking the cut and fold lines as shown. The dotted arrow indicates where it joins the other half of the net (Fig. 5.60).

4. PART 2 (Step 4) – is identical to the first part, except for the tabs – follow carefully (Fig. 5.61).

5. When both parts are made, cut out along the solid lines: score along the dashed lines. Fold upwards along the scored lines.

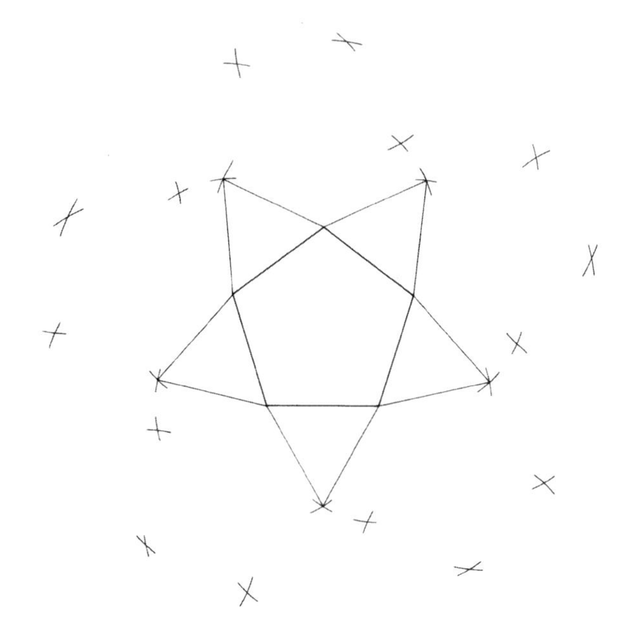

Fig. 5.58. Step 1.

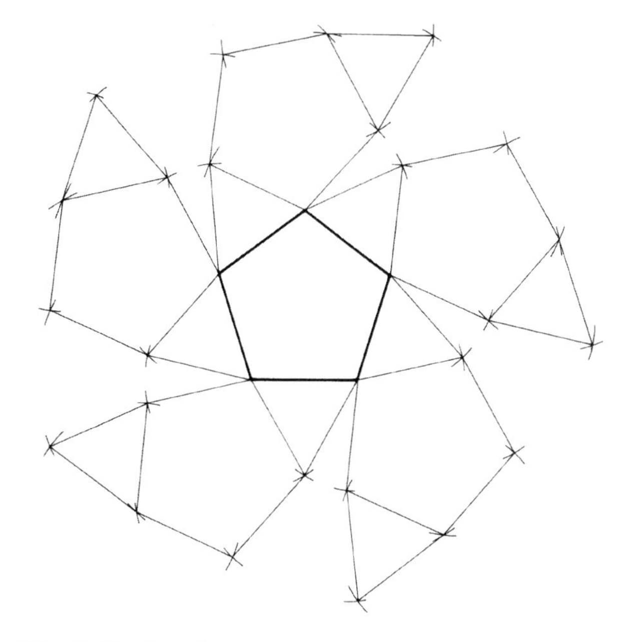

Fig. 5.59. Step 2.

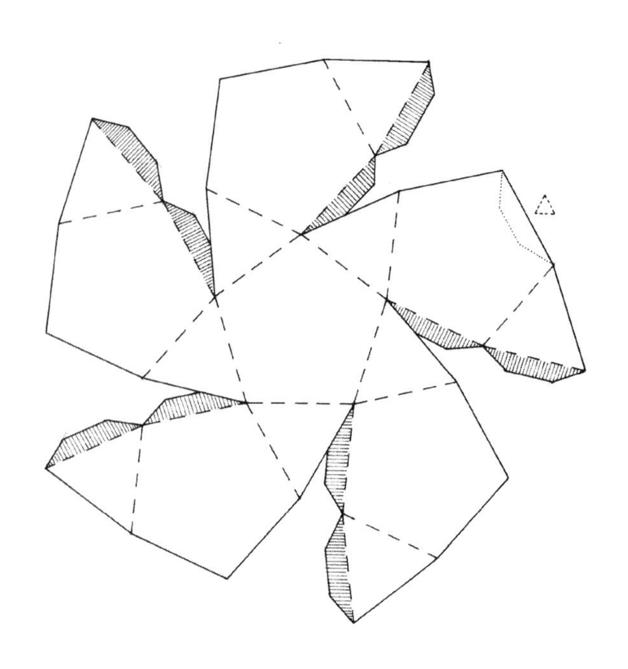

Fig. 5.60. Step 3.

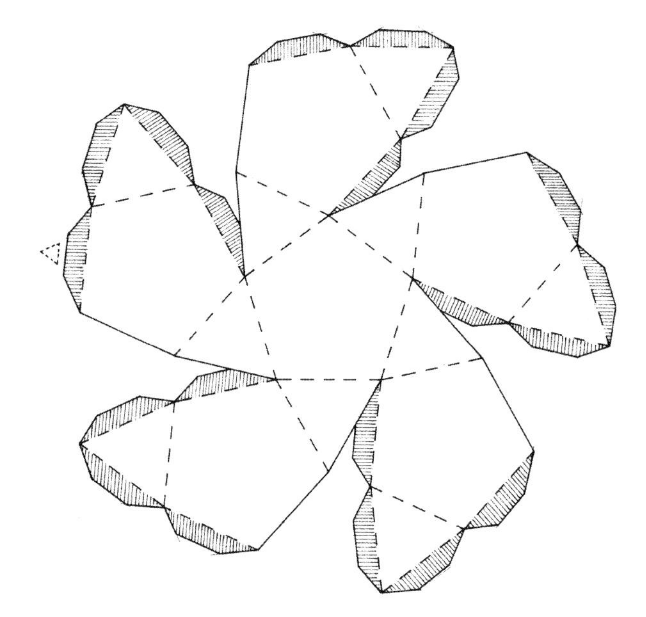

Fig. 5.61. Step 4.

Figs. 5.62–5.63. Illustration of the two halves of the net drawn out on card and cut out.

Fig. 5.64. Illustration of the two halves of the net glued.

6. Glue carefully and progressively (tabs inside) – one or two tabs at a time. Clean up and decorate as you wish.

Fig. 5.65. Illustration of the model complete.

Net for Truncated Dodecahedron

32 faces (20 triangles, 12 decagons), 60 vertices, 90 edges. Icosahedral (2, 3, 5-fold) symmetry.

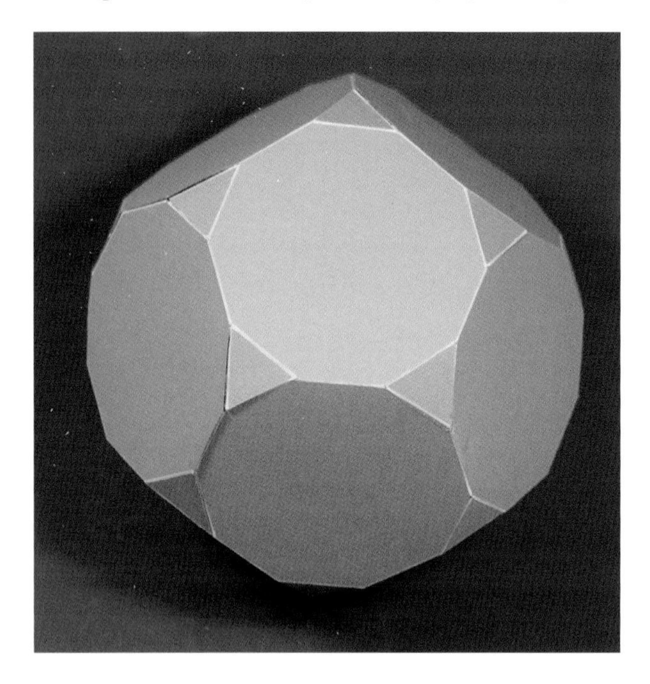

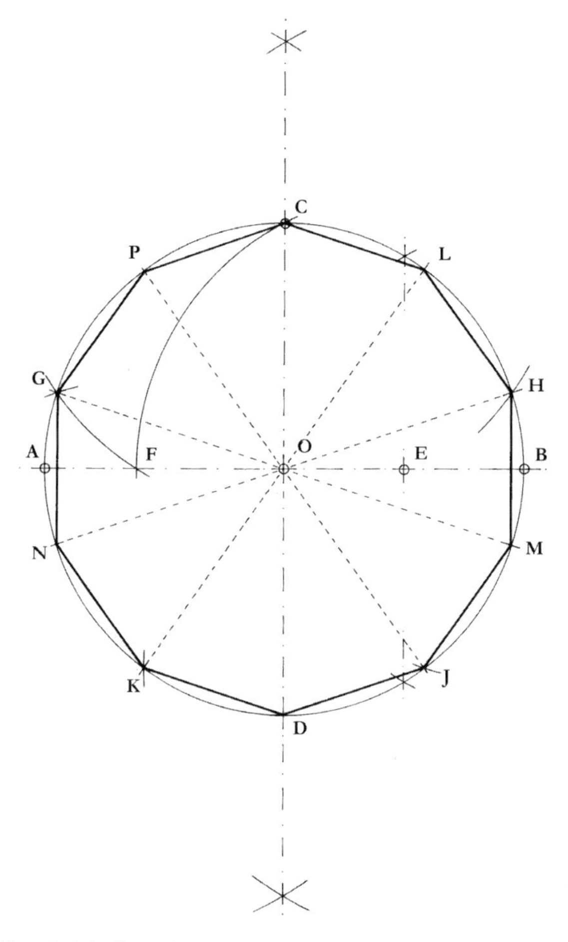

Fig. 5.66. Step 1.

Method:

This net is made in TWO parts.

1. PART 1. (Step 1) To draw the required decagon (ten-sided polygon), use any previous method to first draw a pentagon, or as follows (Fig. 5.66): on horizontal AB draw a circle radius OA. Arc above and below from A and B with radius AB, and connect the intersection points to find the vertical which crosses the circle at C and D. Find the midpoint of OB by drawing arcs from B with radius OB to cross the circle above and below. Join the intersection points to find point E. With compass on E and radius EC draw arc down to cross the horizontal at F. With compass on C and radius CF swing an arc to both sides to give points G and H. Walk this radius round to find the other two points of pentagon CHJKG. Draw lines from each pentagon point through the centre to the circle on the other side to find the rest of the points of the decagon CLHMJDKNGP.

2. (Step 2) Add further decagons to alternate sides of the first decagon by first drawing arcs from each pair of points on the central decagon in turn with the same radius as that of the central circumcircle (the circle that encircles the decagon) to find the centres of the circumcircles of the surrounding decagons.

3. Then walk the side length round each circle to mark the points of each decagon (Fig. 5.67).

4. (Step 3) Tune these points by drawing lines through existing and new points. A selection of checking lines only is shown here (Fig. 5.68).

5. (Step 4) Add equilateral triangles to the faces shown in (Fig. 5.69. See also Fig. 5.15, page 49).

6. (Step 5) Add tabs and cut and fold lines as shown. The dotted arrow and tab indicates where the other part of the net joins (Fig. 5.70).

7. PART 2 (Step 6) – is identical to the first part, except for the tabs – follow carefully (Fig. 5.71).

8. When both parts are made, cut out along the solid lines: score along the dashed lines. Fold upwards along the scored lines.

9. Glue carefully and progressively (tabs inside) – one or two tabs at a time. Clean up and decorate as you wish.

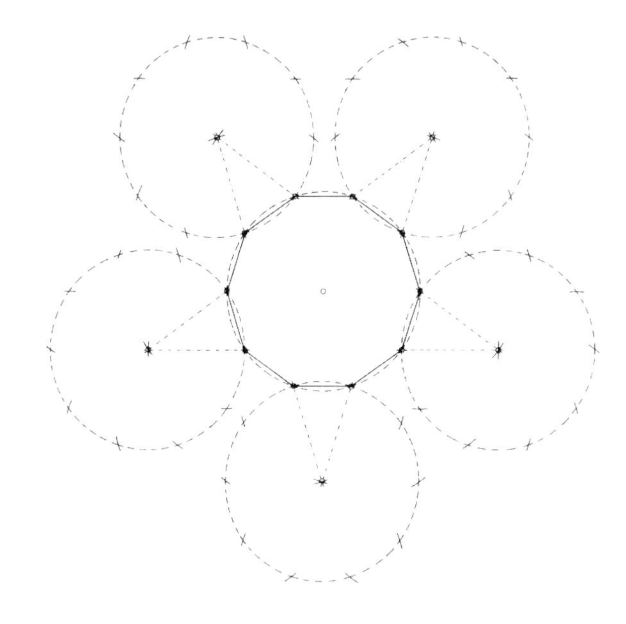

Fig. 5.67. Step 2.

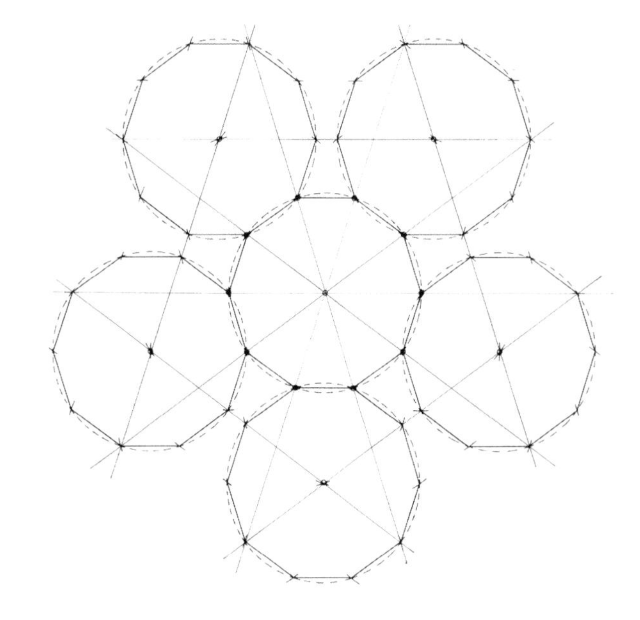

Fig. 5.68. Step 3.

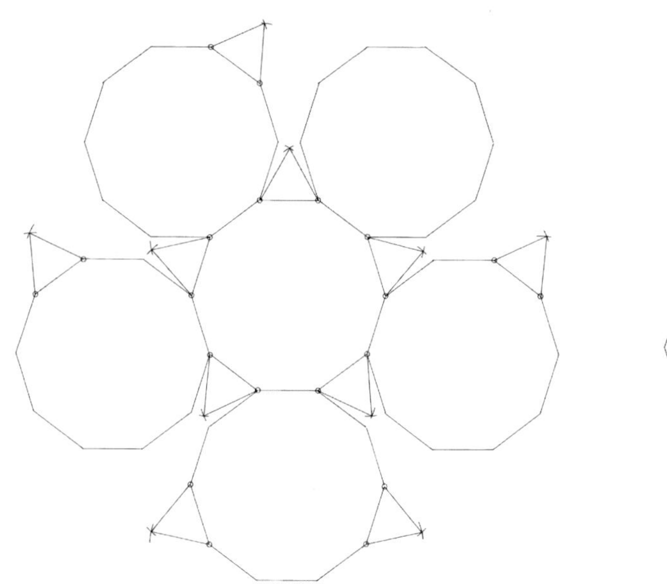

Fig. 5.69. Step 4.

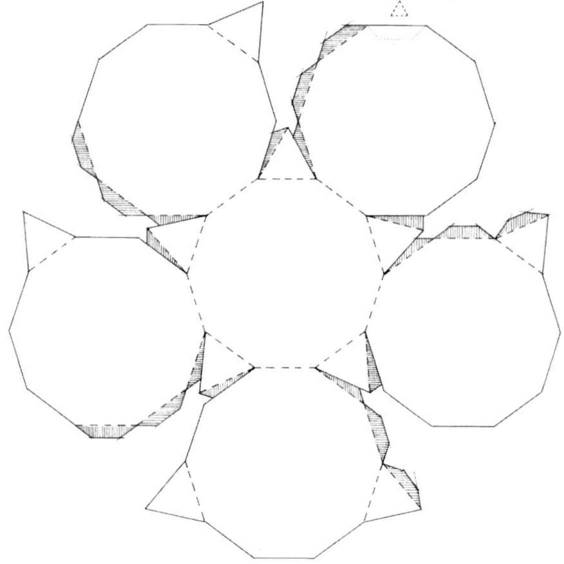

Fig. 5.70. Step 5.

Fig. 5.71. Step 6.

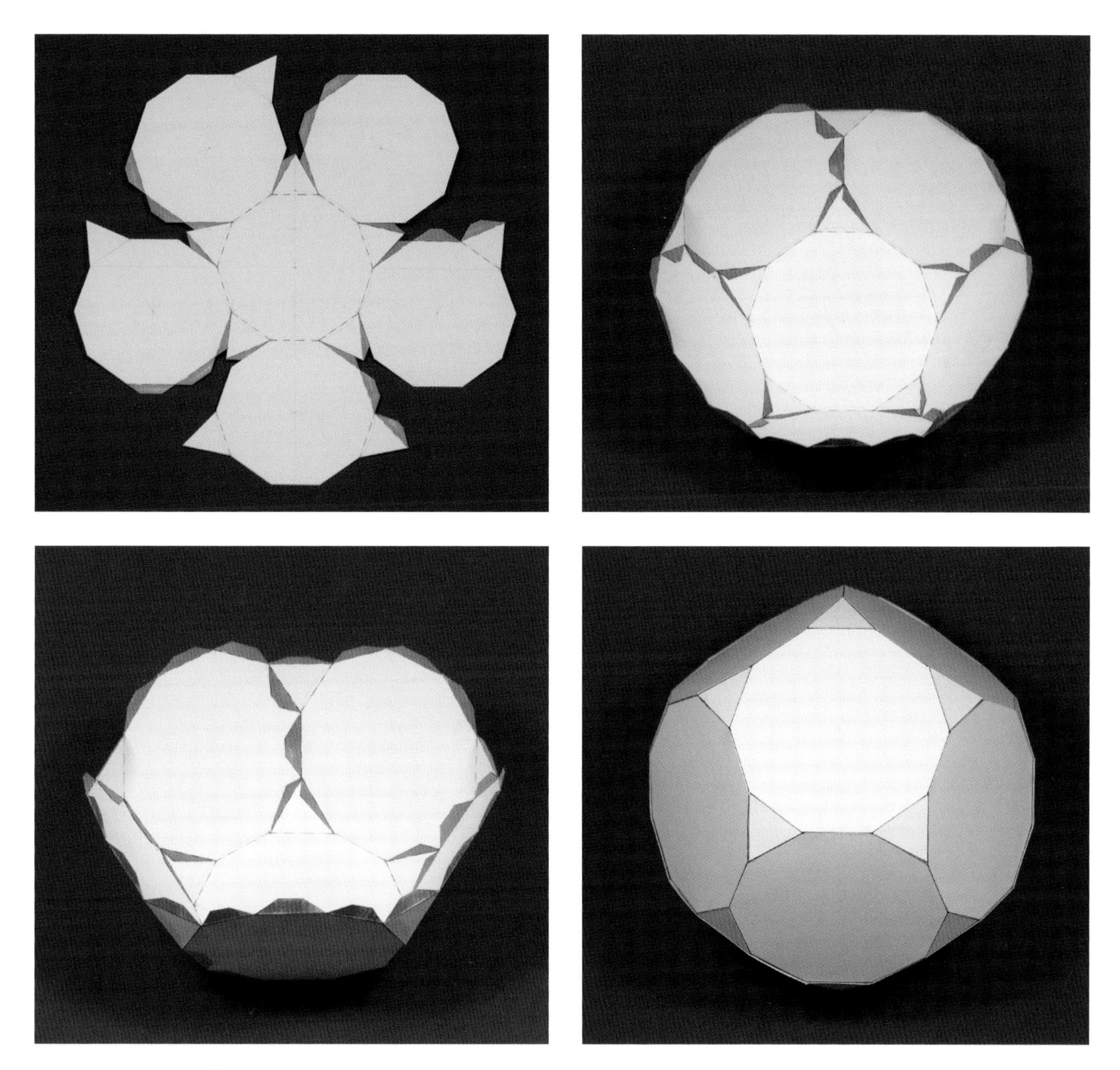

Figs. 5.72–5.75. Illustrations of one half of the net drawn out on card and cut out (top left); the two halves assembled (top right and bottom left); and the model complete (bottom right).

Net for Snub Cube

38 faces (32 triangles, 6 squares), 24 vertices, 60 edges. Octahedral (2, 3, 4-fold) symmetry.

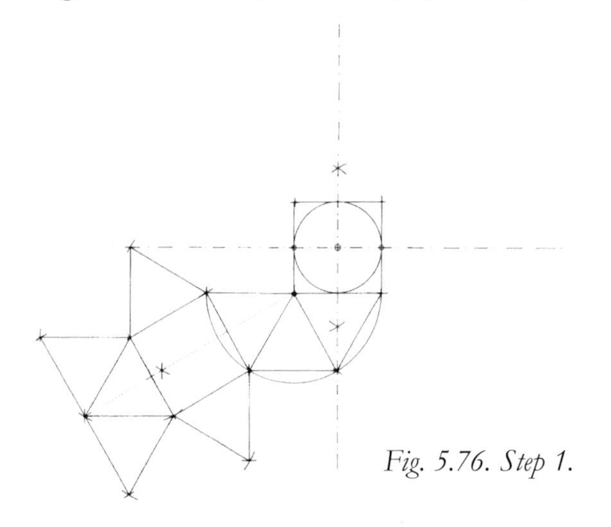

Fig. 5.76. Step 1.

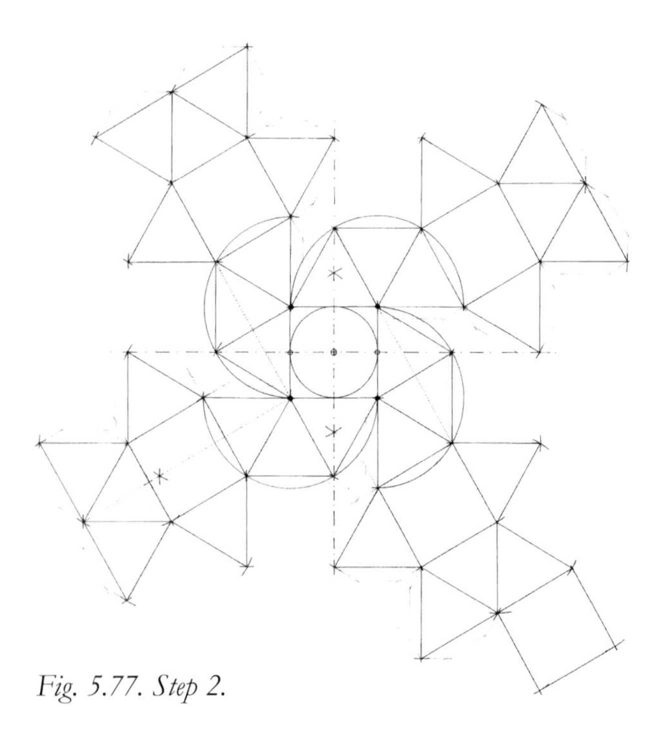

Fig. 5.77. Step 2.

Method:

1. (Step 1) On a horizontal draw a circle in the middle of your paper and add the vertical axis. About this circle construct a square.

2. About the lower lefthand corner of the square draw a semicircle radius the side length of the square. Extend the lower side of the square to intersect the semicircle. Step the radius round the semicircle, and join the intersection points to each other and back to the centre of the semicircle to create three equilateral triangles.

3. Construct a square on the side of one triangle as shown; and on each side of the square construct triangles. Draw triangles on the two free sides of the outer triangle (Fig. 5.76).

4. (Step 2) On each of the other three sides of the first square, carry out the same construction: except that the righthand leg has an additional square on the end (Fig. 5.77).

5. (Step 3) Add tabs and mark the cut and fold lines as shown (Fig. 5.78).

6. Cut out along the solid lines: score along the dashed lines. Fold upwards along the scored lines. Glue carefully and progressively (tabs inside) – one or two tabs at a time. Clean up and decorate as you wish.

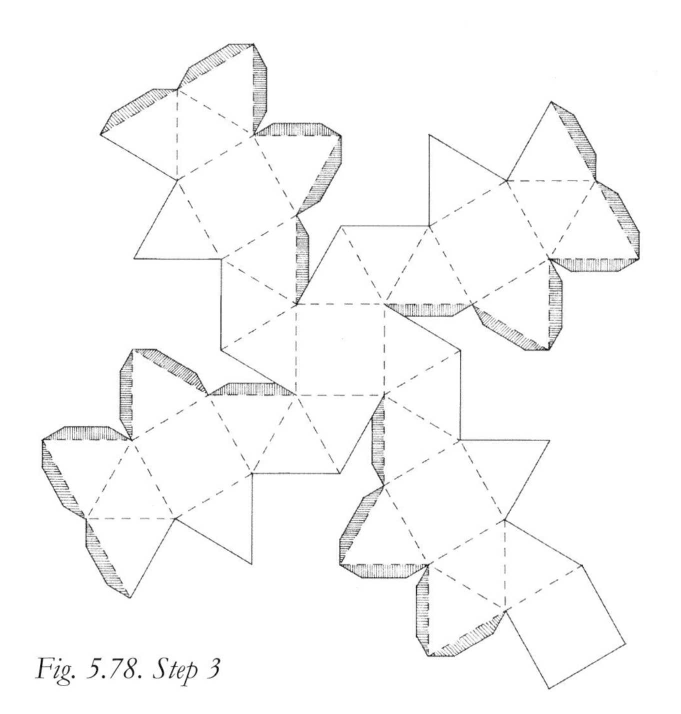

Fig. 5.78. Step 3

Figs. 5.79–5.80. Illustrations of the figure drawn out on card and cut out (top); and partially assembled (bottom).

Figs. 5.81–5.83. Further steps in the assembly (top left and right); and the model complete (bottom right).

Net for Truncated Icosidodecahedron

62 faces (30 squares, 20 hexagons, 12 decagons), 120 vertices, 180 edges. Icosahedral (2, 3, 5-fold) symmetry.

Method:

This net is made in SIX parts

1. PART 1. First draw a decagon of the required side length (see page 74 for method).

2. (Step 1) On alternate sides of the decagon, construct squares by projecting lines from the opposite corners through and beyond the face on which the square is to be constructed. Then draw arcs with radius the side length to establish the outer limit of the square – (only one pair of projected lines is shown here for clarity).

3. (Step 2) On one side of each square, construct a hexagon by first finding the centre of the circumcircle of the hexagon by arcing from each end of the side in turn, drawing the circle with the same radius, and then walking round the circle to find the other points of the hexagon (Fig. 5.85, see also Fig. 5.43, page 65).

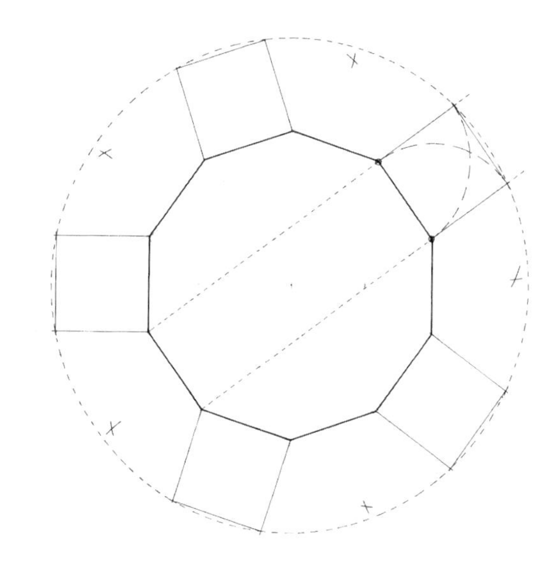

Fig. 5.84. Step 1.

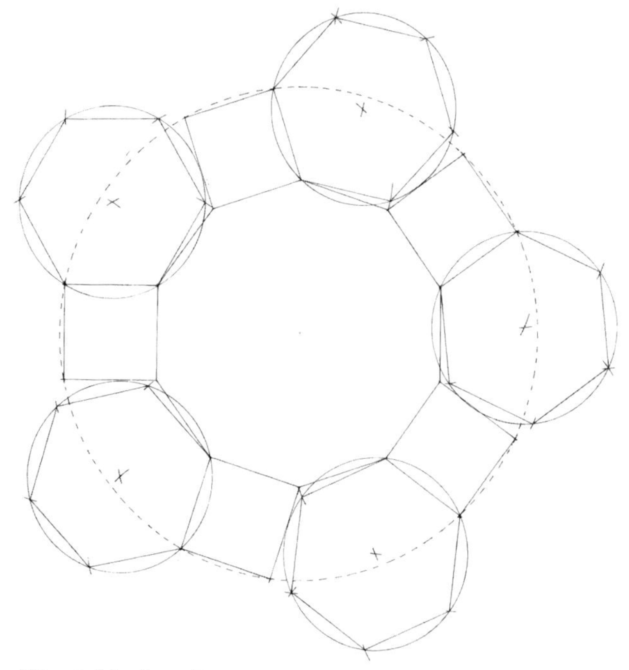

Fig. 5.85. Step 2.

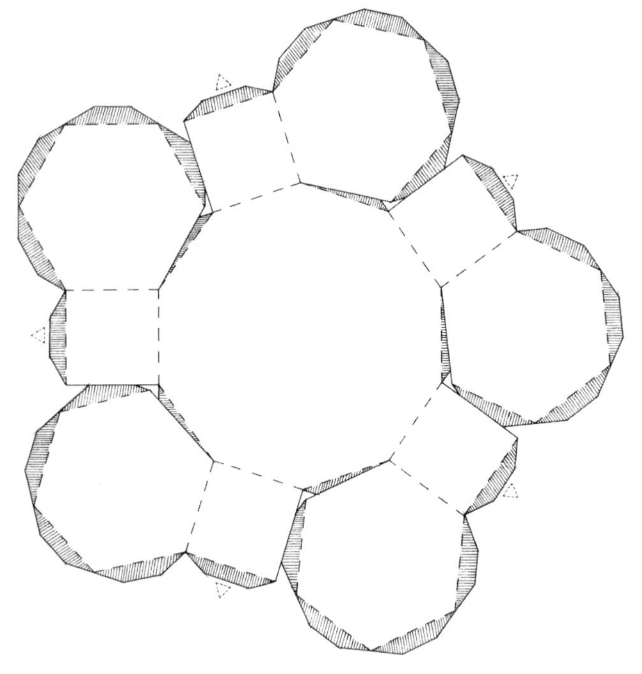

Fig. 5.86. Step 3.

4. (Step 3) Complete the first part of the net by marking the cut and fold lines, and adding tabs, as shown. The dotted arrows indicate where the other five parts of the net join (Fig. 5.86).

5. PARTS 2–5 (Step 4) Start with a decagon of the same size as above, and construct squares and hexagons on the sides as shown using the same techniques. Add a second decagon on the right, with a square and hexagon as shown (Fig. 5.87).

6. (Step 5) Complete by adding tabs and marking the cut and fold lines as shown. The dotted arrow indicates where these four parts join the central part of the net (Fig. 5.88).
   You will need FOUR of these.

7. PART 6 (Step 6) is similar to the previous ones, but with an extra decagon on one end (Fig. 5.89).

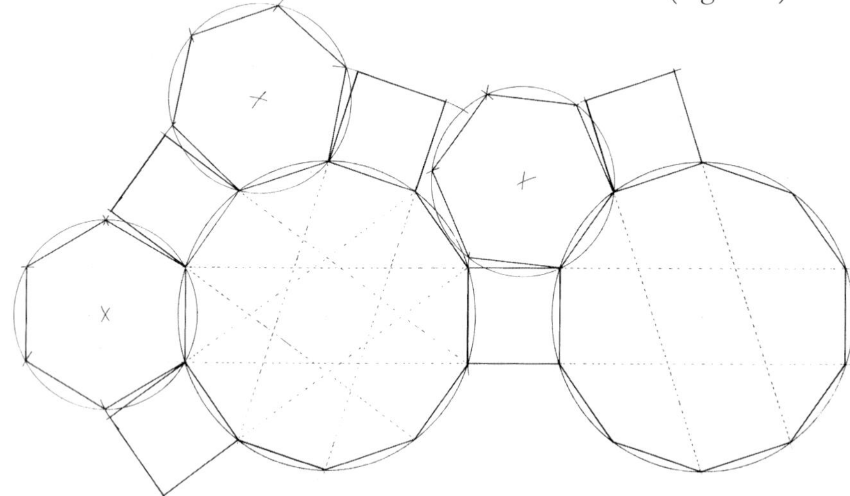

Fig. 5.87. Step 4.

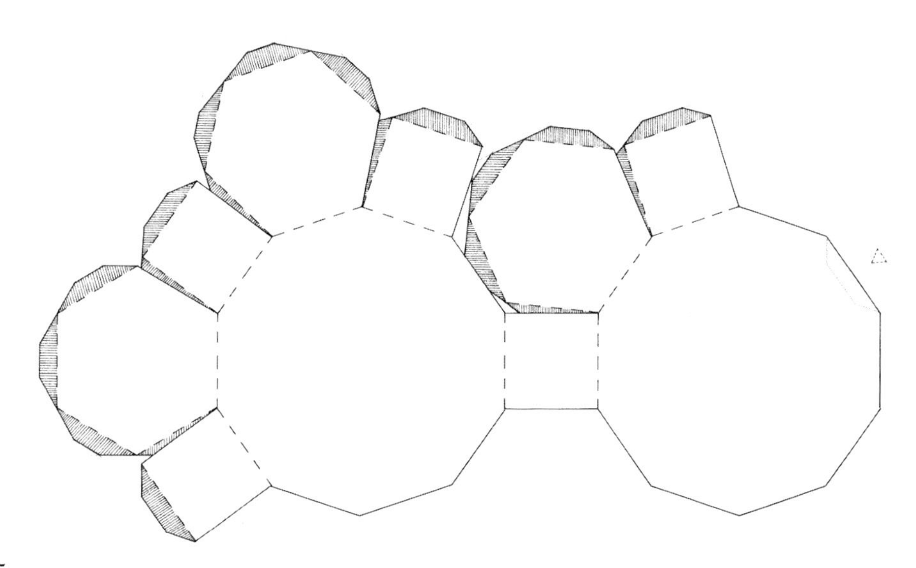

Fig. 5.88. Step 5.

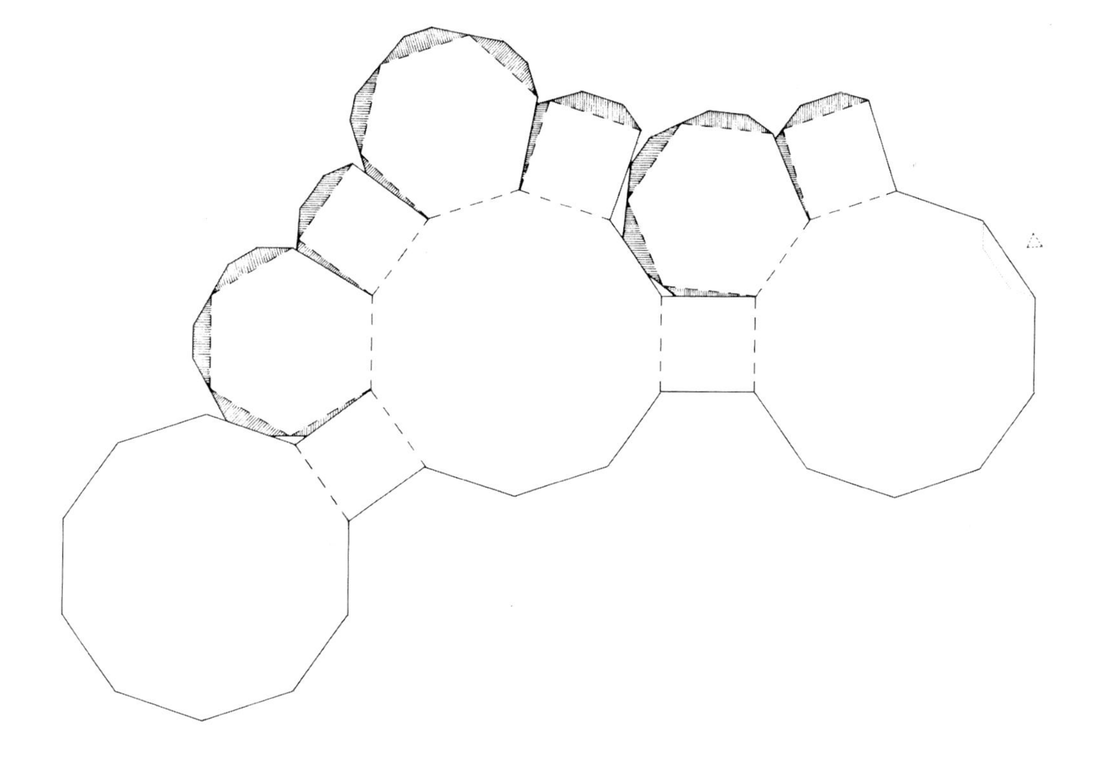

Fig. 5.89. Step 6.

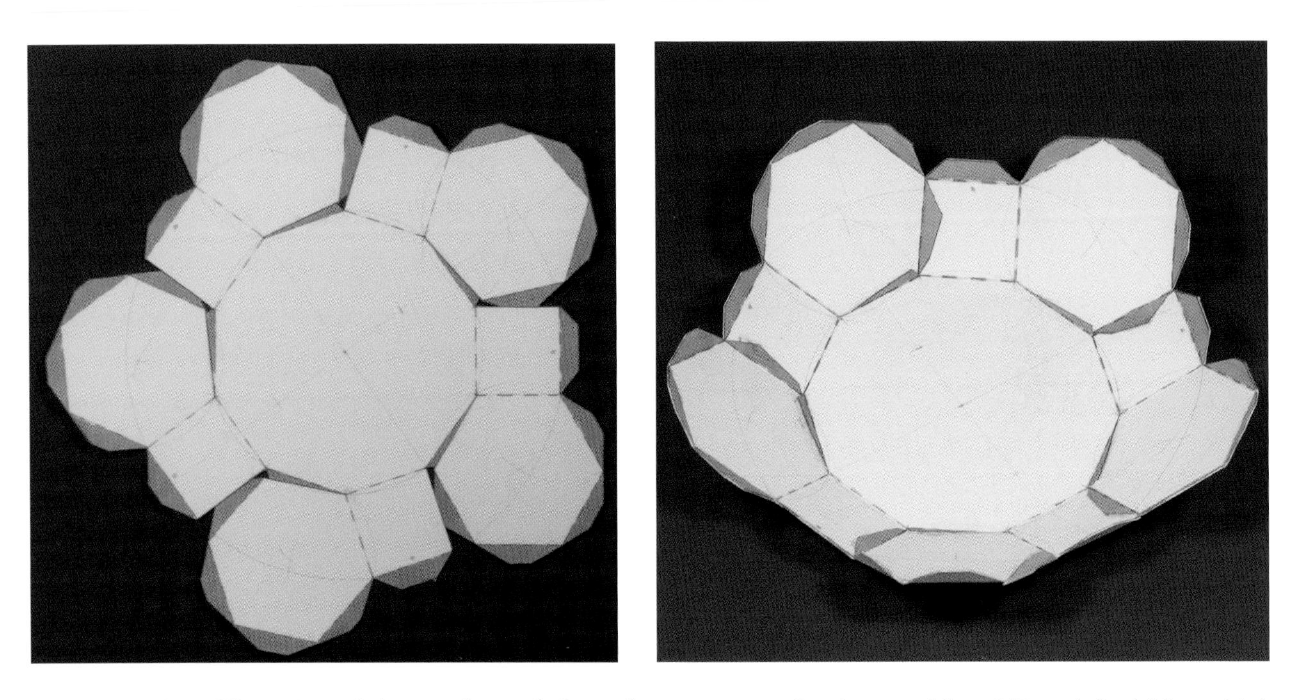

Figs. 5.90–5.91. Illustrations of the central part of the net drawn out on card and cut out (above left); and glued (above right).

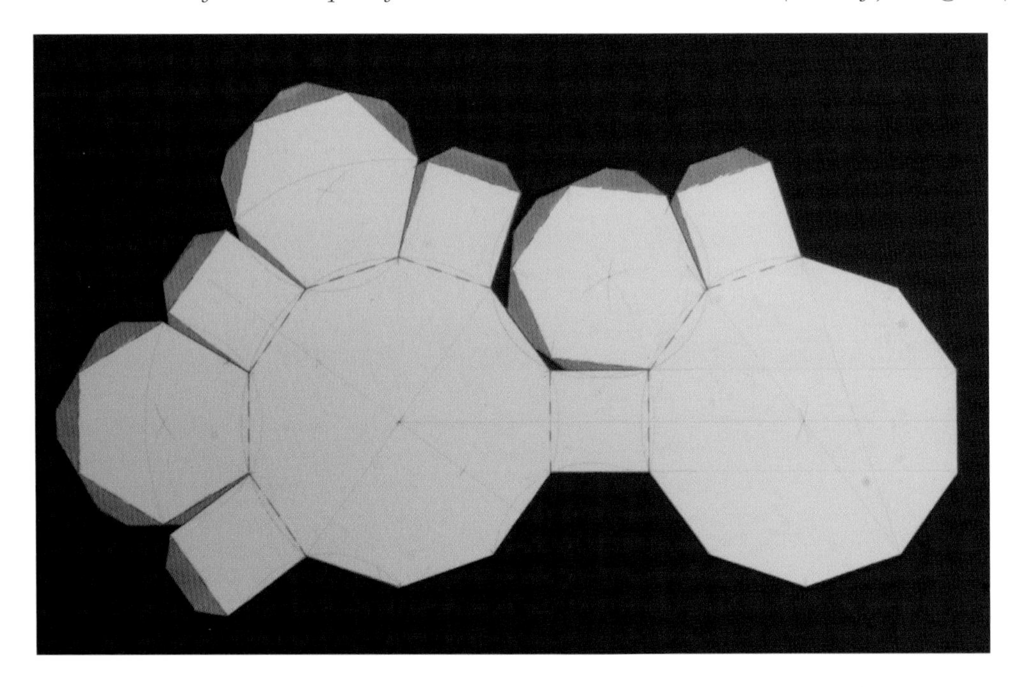

Fig. 5.92. Illustration of the one of the four parts of the net drawn out on card and cut out.

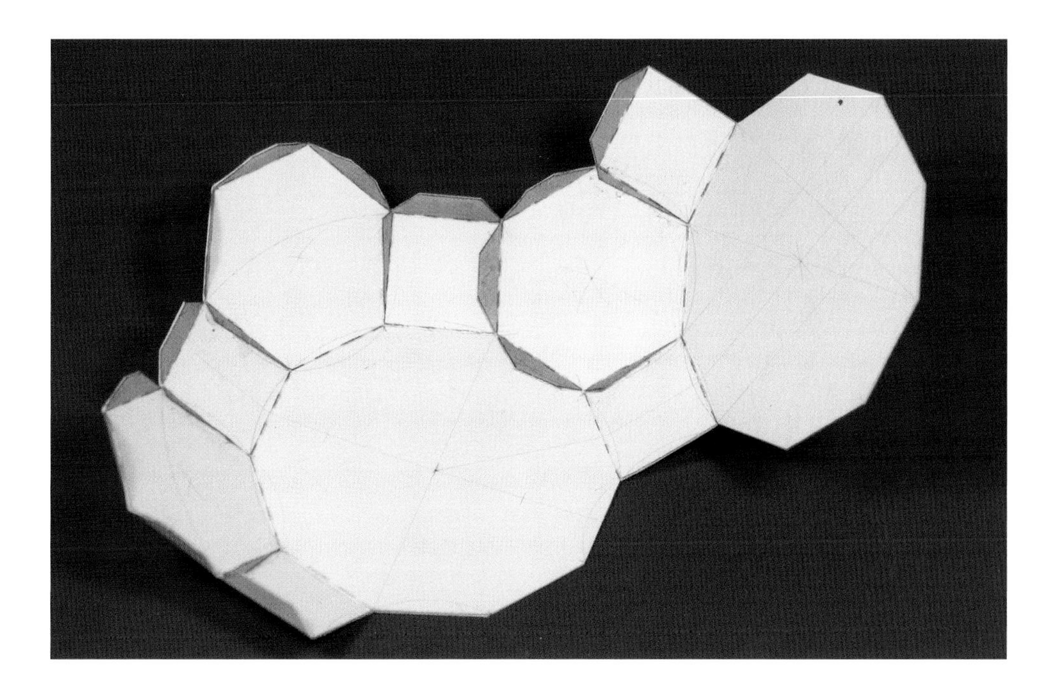

Fig. 5.93. Illustration of the one of the four parts of the net drawn out on card, cut out and glued.

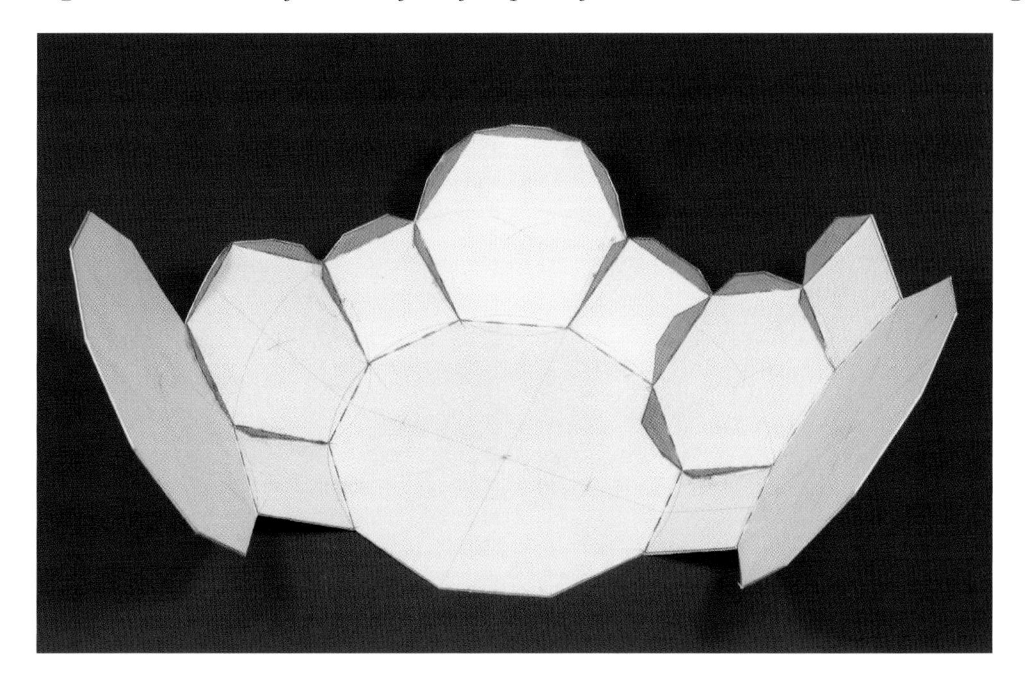

Fig. 5.94. Illustration of the sixth part of the net drawn out on card, cut out and glued.

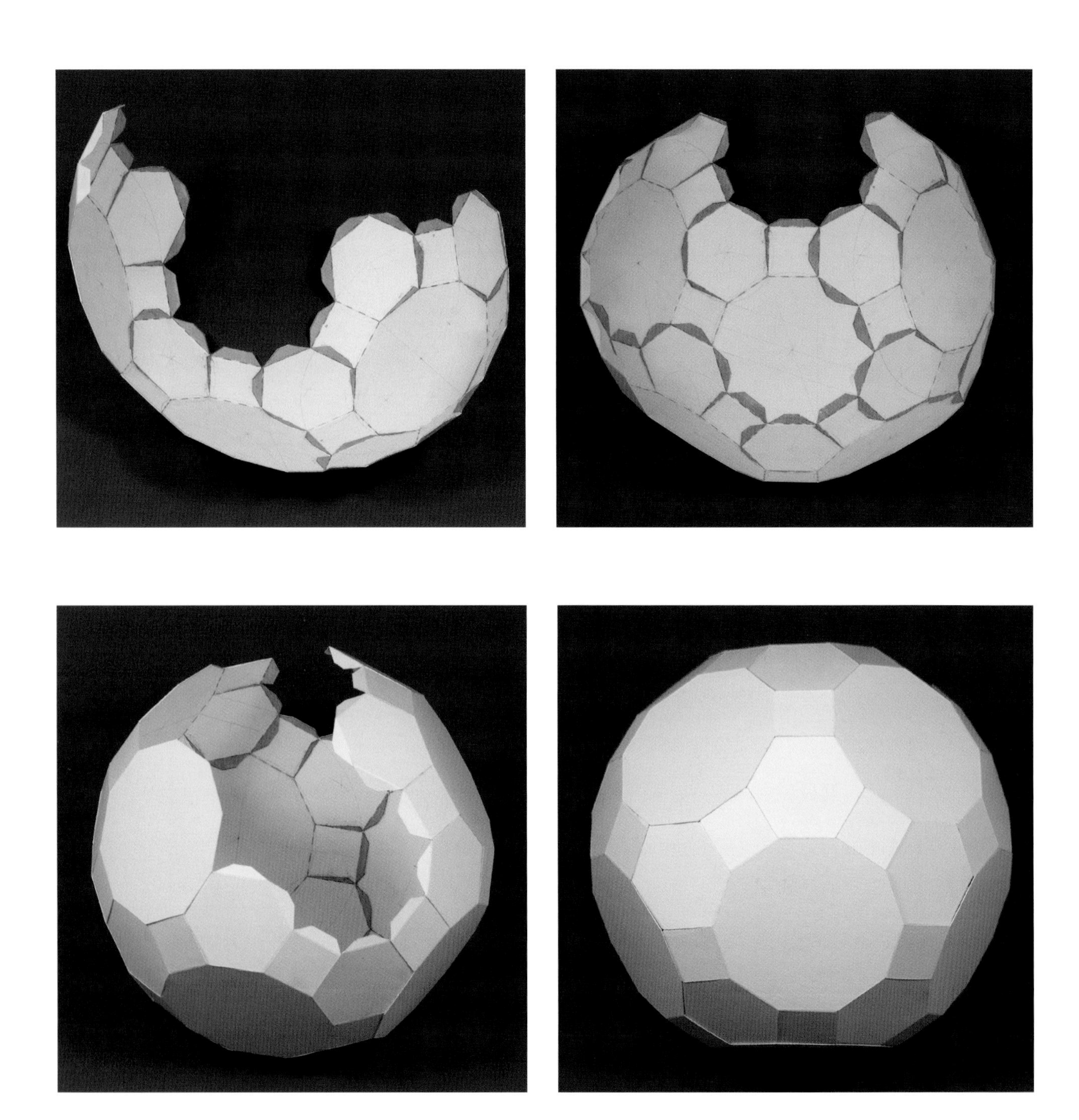

Figs. 5.95–5.98. Illustrations of the model being assembled in stages, and complete (bottom right).

Net for Rhombicosidodecahedron

62 faces (20 triangles, 30 squares, 12 pentagons), 60 vertices, 120 edges. Icosahedral (2, 3, 5-fold) symmetry.

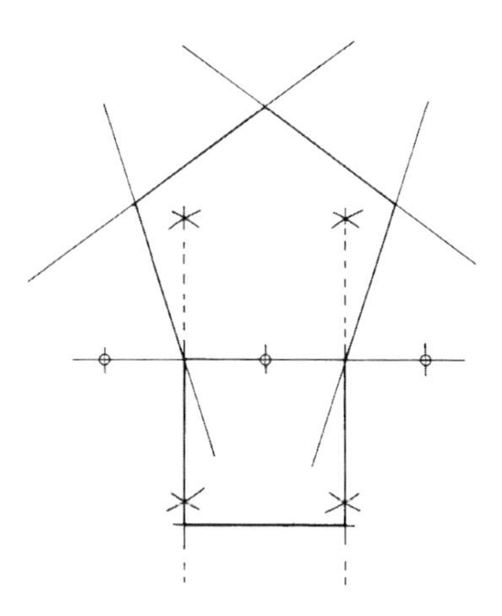

Fig. 5.99. Step 1

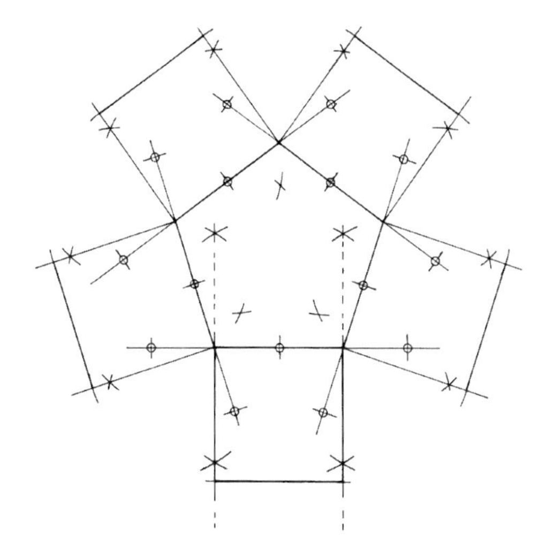

Fig. 5.100. Step 2.

Method:

This net is made in SIX parts.

1. PART 1 (Steps 1–3) Start with a pentagon of the required side length (see page 67 for method). Add a square to each face (see Figure 5.15, page 49), and equilateral triangles to the faces shown (see Figure 5.16, page 49). Add tabs plus cut and fold lines as shown. The dotted arrows indicate where the other five parts of the net join (Figs. 5.99–5.101).

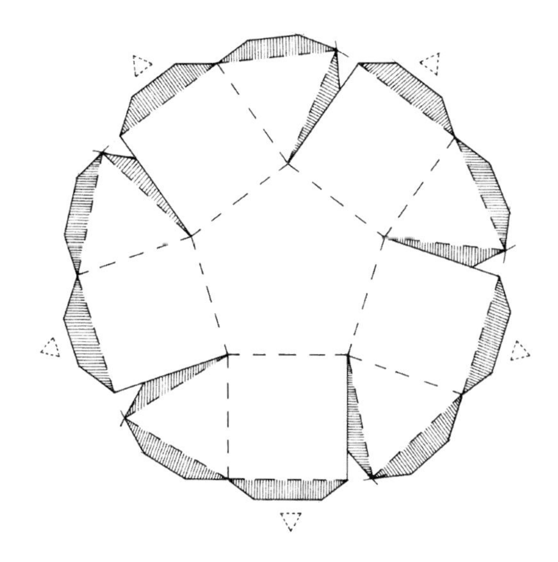

Fig. 5.101. Step 3.

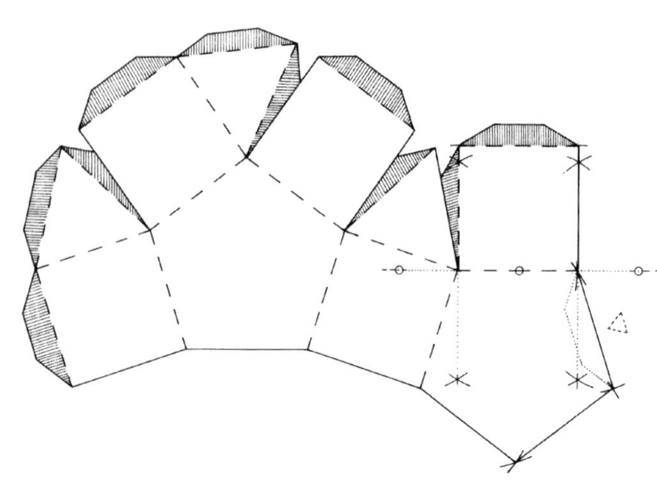

Fig. 5.102. Step 4.

2. PARTS 2–5 (Step 4) Starting again with a pentagon with the same edge length as used in the first drawing, add triangles and a further pentagon with triangles as shown. To draw the second pentagon follow Fig. 5.51 on page 68. Add tabs plus cut and fold lines as shown. The dotted arrow indicates where these four parts join the central part of the net (Fig. 5.102). You will need FOUR of these.

3. PART 6 (Step 5) – is similar to the previous ones, but with an extra pentagon on the left (Fig. 5.103).

4. When all six parts are made, cut out along the solid lines: score along the dashed lines. Fold upwards along the scored lines.

5. Glue carefully and progressively (tabs inside) – one or two tabs at a time. Clean up and decorate as you wish.

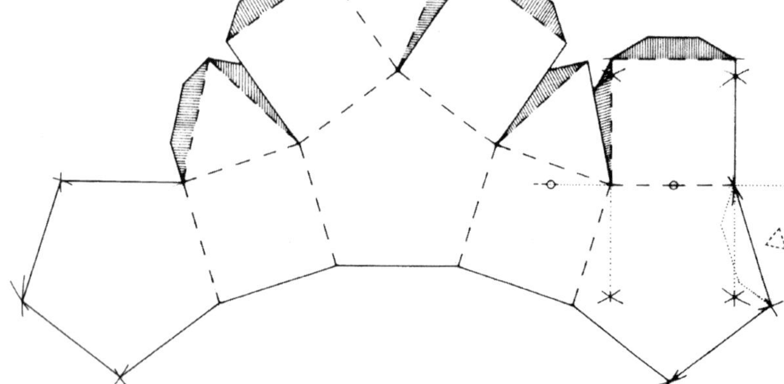

Fig. 5.103. Step 5.

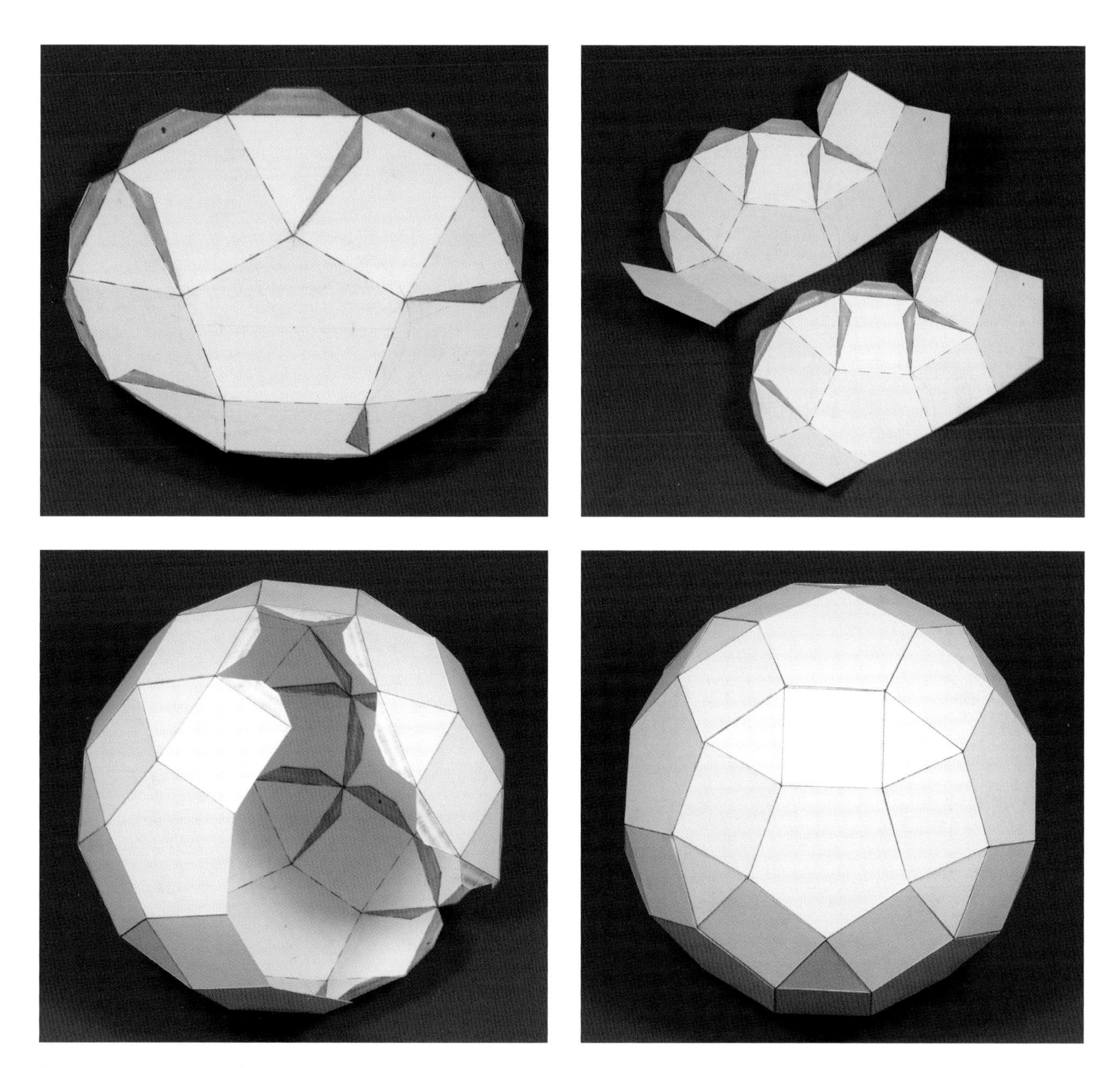

Figs. 5.104–5.107. Illustrations of the model being assembled in stages; and complete (bottom right).

Net for Snub Dodecahedron

92 faces (80 triangles, 12 pentagons), 60 vertices, 150 edges. Icosahedral (2, 3, 5-fold) symmetry.

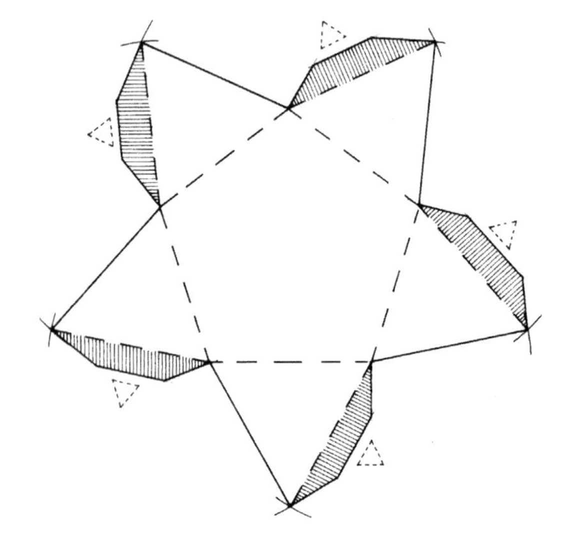

Fig. 5.108. Step 1.

Method:

This net is made in SIX parts

1. PART 1 (Step 1) Start with a pentagon with the required side length (see page 67). Add an equilateral triangle to each side, and tabs plus cut and fold lines as shown. The dotted arrows indicate where the other five parts of the net join (Fig. 5.108).

2. PARTS 2–5 (Step 2) Starting again with a pentagon with the same edge length as used in the first drawing, add triangles and a further pentagon with triangles as shown. To draw the second pentagon follow Figure 5.51 on page 68. Add tabs plus cut and fold lines as shown. The dotted arrow indicates where these four parts join the central part of the net (Fig. 5.109). You will need FOUR of these.

3. PART 6 (Step 3) is similar to the previous ones, but with an extra pentagon on the top left (Fig. 5.110).

4. When all six parts are made, cut out along the solid lines: score along the dashed lines. Fold upwards along the scored lines.

5. Glue carefully and progressively (tabs inside) – one or two tabs at a time. Clean up and decorate as you wish.

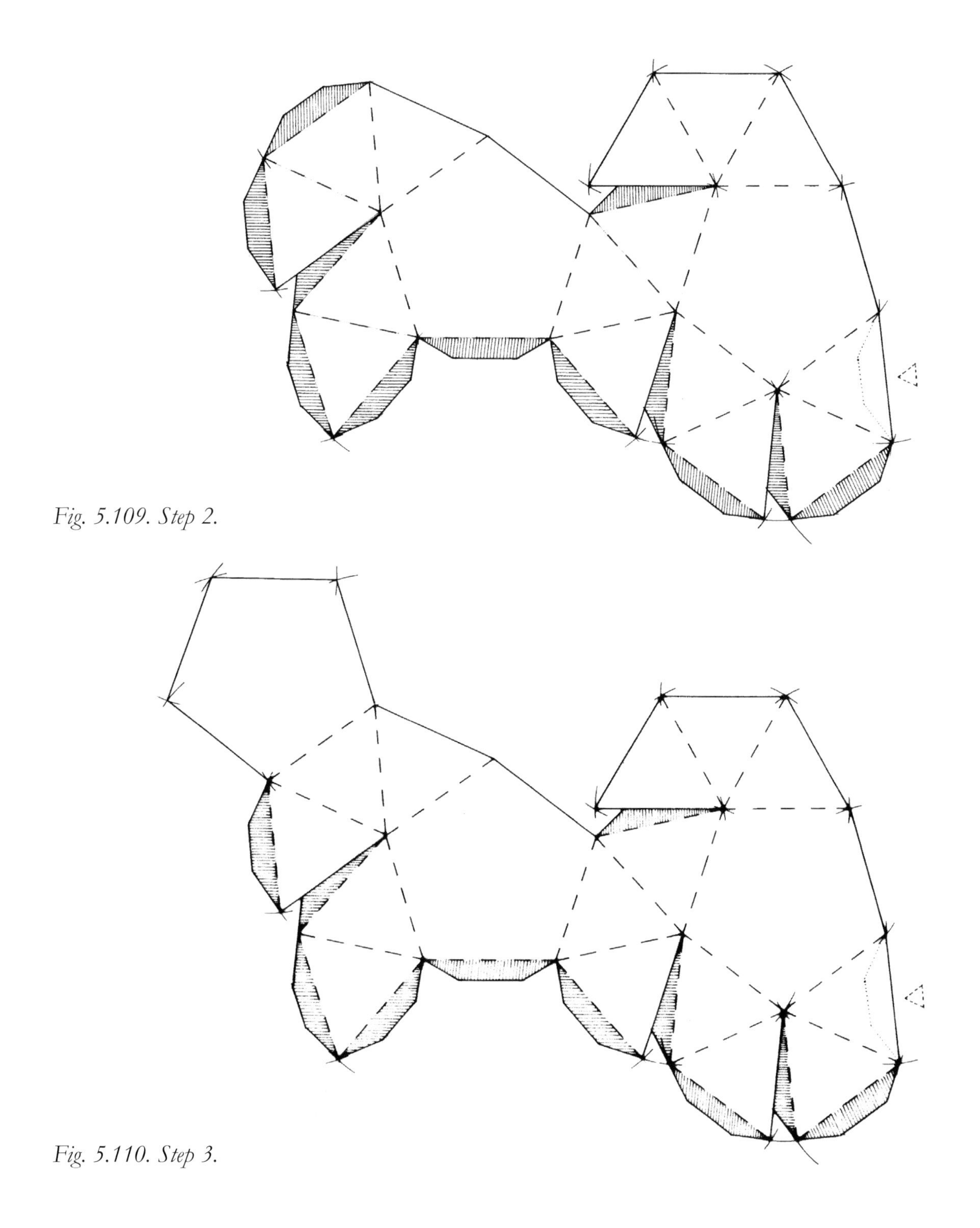

Fig. 5.109. Step 2.

Fig. 5.110. Step 3.

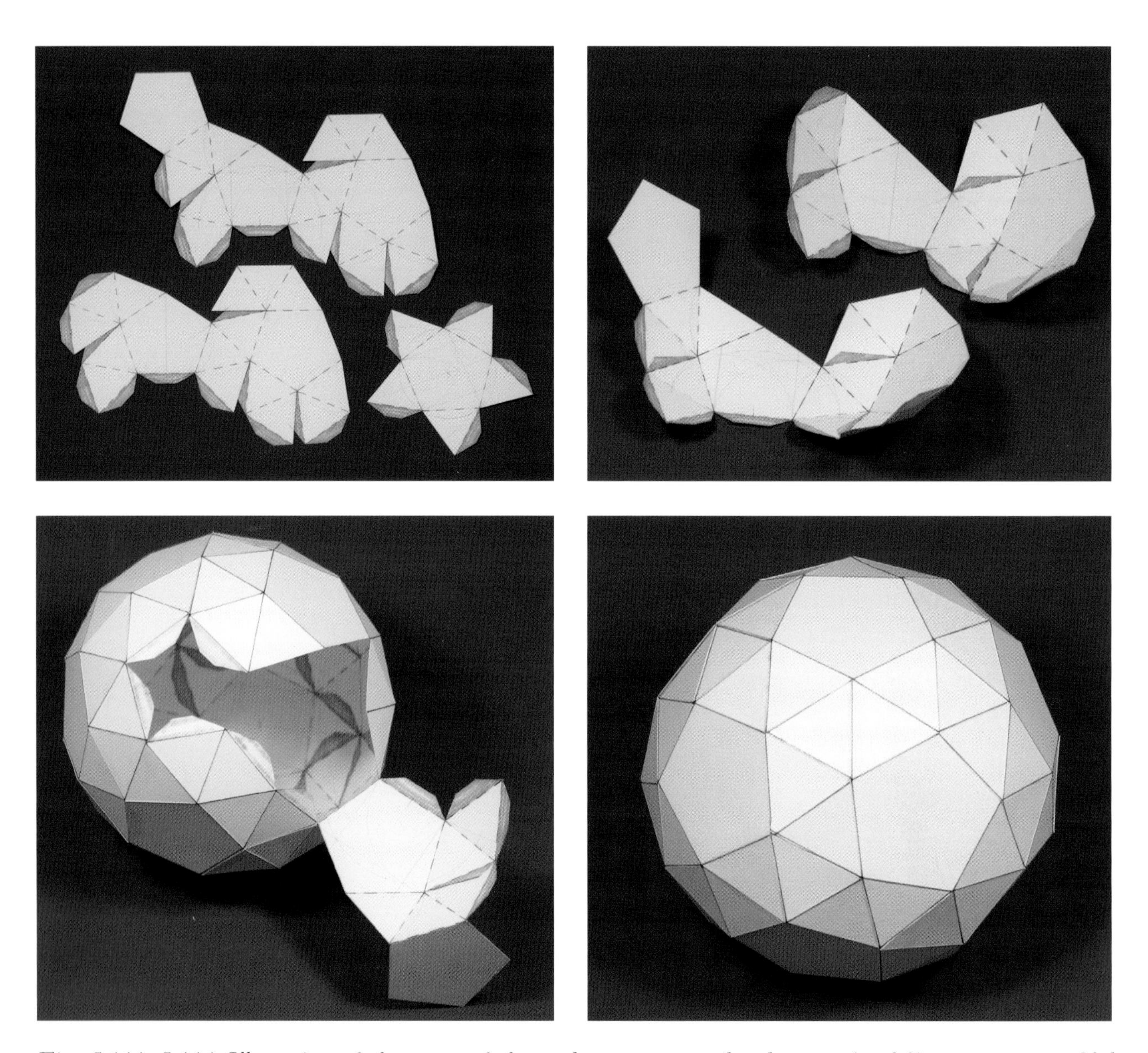

Figs. 5.111–5.114. Illustrations of three parts of the net drawn out on card and cut out (top left); two parts assembled (top right); the model partially assembled (bottom left); and complete (bottom right).

Paperfolds for Other Polyhedra

Fig. 6.1. Models of the Compound Solids of Icosahedron + Dodecahedron (top); Cube + Octahedron (left); and two Tetrahedra (right).

Platonic Compound: Two Tetrahedra

Method:

1. Draw out a net for a tetrahedron (see page 30). Add the half-sized triangles as shown in Figure 6.2 (left). Pounce (pierce through the paper with a compass point) the intersection points through so that the same pattern can be drawn on the visible faces of the model. Then assemble.

2. Draw out nets for a further FOUR tetrahedra with edge lengths exactly half that of the first, and assemble. It can make a neater model if the top triangle of the net is omitted as in Figure 6.2 (right).

3. Stick the four smaller tetrahedra onto the faces of the larger in the positions shown dotted to create the compound solid of two interpenetrating Tetrahedra (see figures 6.3 and 6.4 opposite).

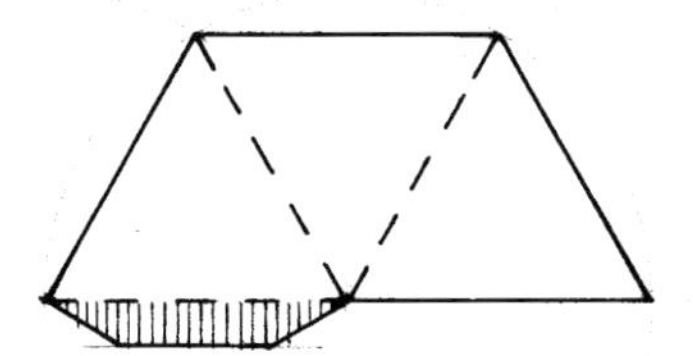

Fig. 6.2. Nets for the core solid (left) and added pyramids (right).

An alternative method is to make an octahedron plus eight tetrahedra with the same edge length, and stick one tetrahedron to each face of the octahedron (which disappears as the core within the resulting solid).

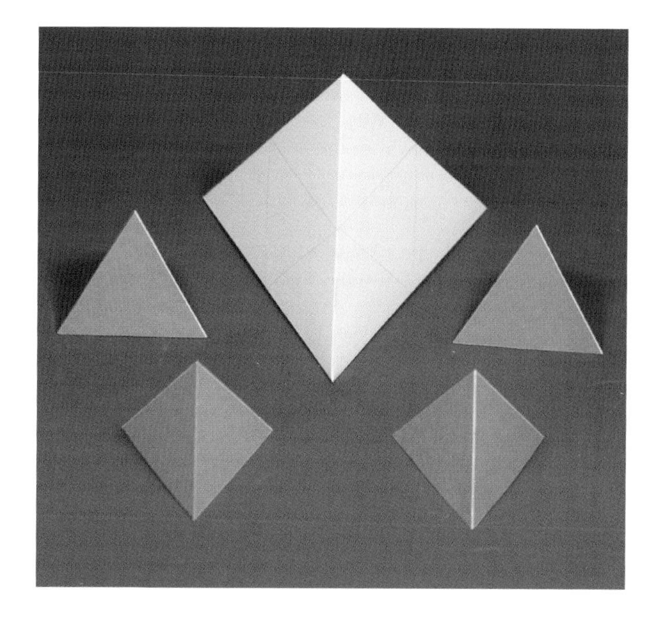

Figs. 6.3–6.4. Illustrations of the described method before and after assembly.

Platonic Compound: Cube + Octahedron

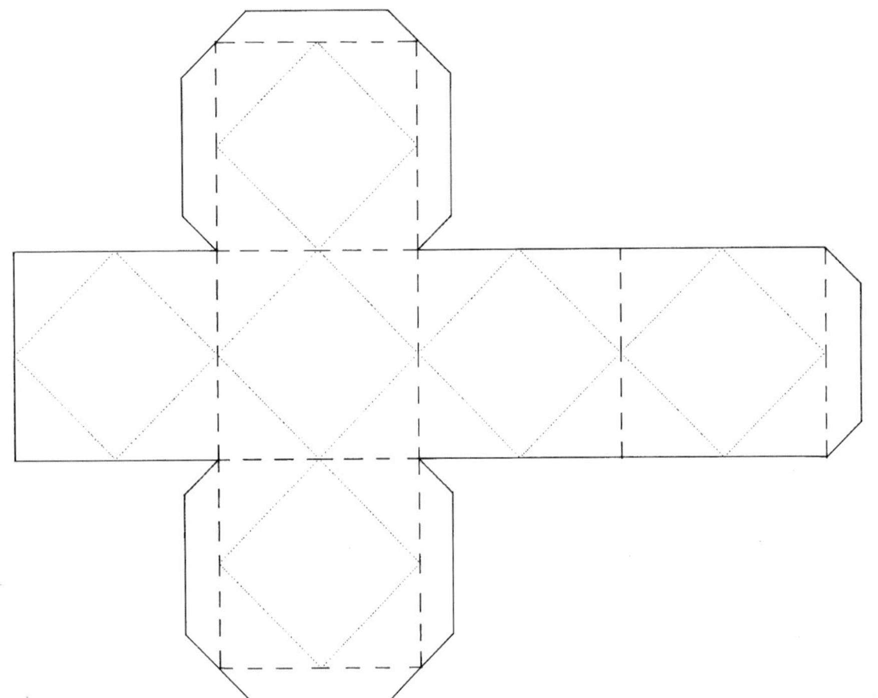

Fig. 6.5. Construction drawings of the net for the Cube.

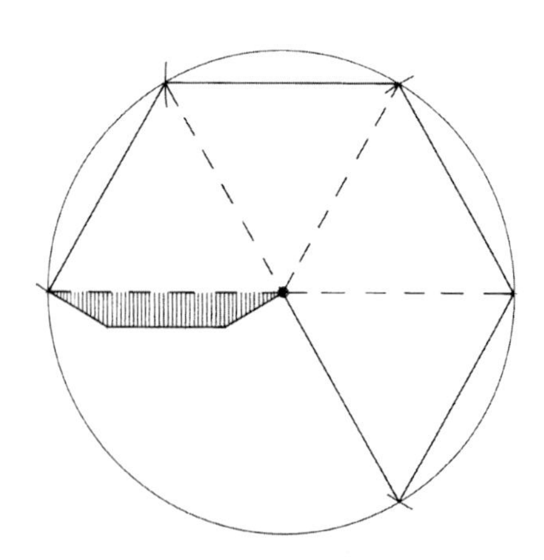

Fig. 6.6. The six half Octahedra.

Method:

1. Draw out a net for a cube (see page 36 and Fig. 6.5) and assemble.

2. Draw out nets for SIX half octahedra (Fig. 6.6) with edge lengths $\sqrt{2}/2$ times the length of the cube's, and assemble.

3. Stick the six half octahedra onto the faces of the cube in the positions shown dotted on the drawing above to create the compound solid of Cube plus Octahedron.

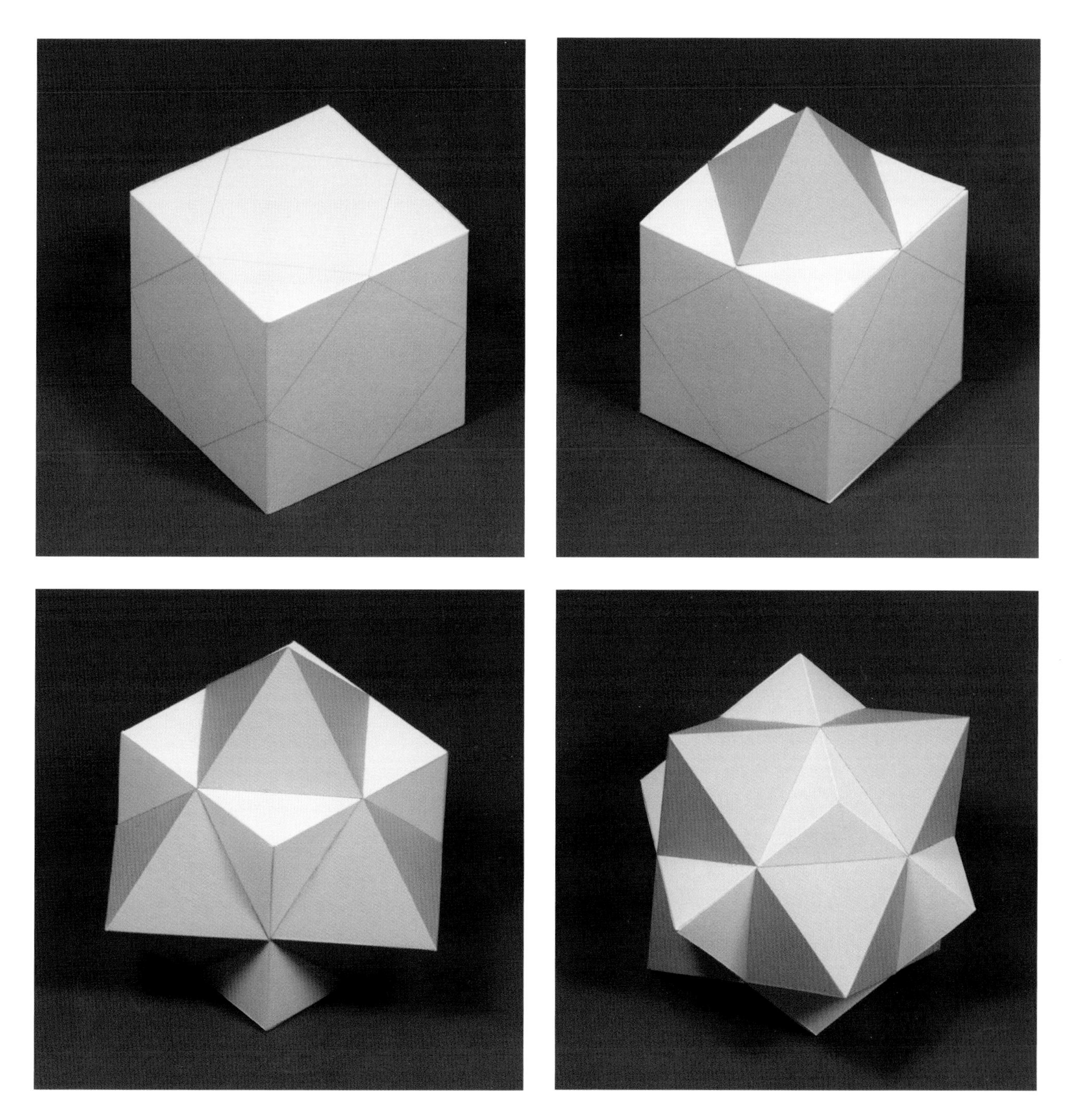

Figs. 6.7–6.10 From top left to bottom right: illustrations of the parent cube; the first half octahedron applied; several more; and the completed model.

Platonic Compound: Icosahedron + Dodecahedron

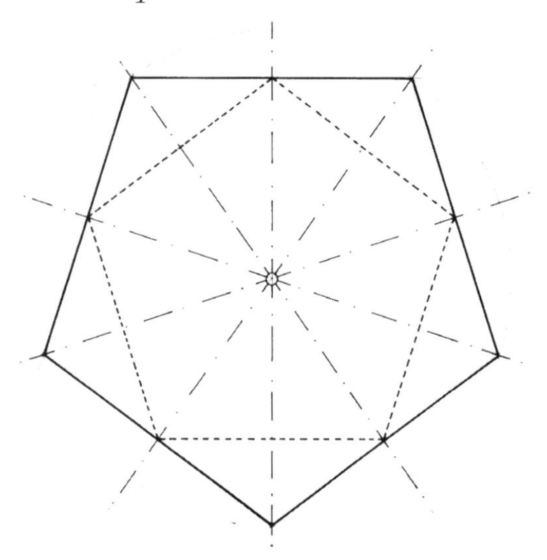

Fig. 6.11. Construction drawings showing the relationship between the dodecahedron faces – drawn in solid line – and the pyramid edges – drawn in broken line.

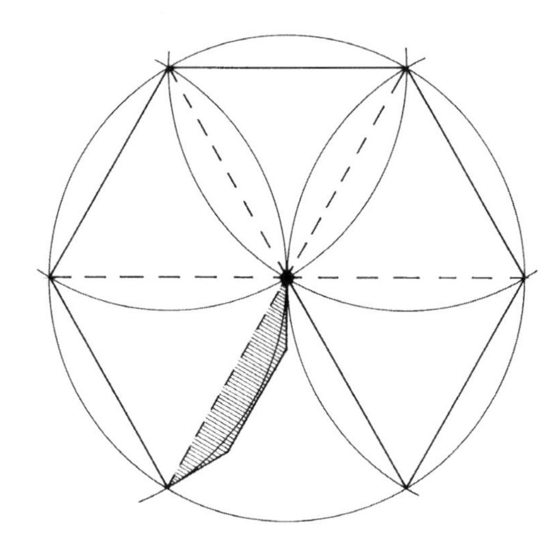

Fig. 6.12. The net of the pyramids drawn using the edge length from the lefthand drawing as the radius and step-round arcs.

Fig. 6.13. John Maine RA included a beautiful partially truncated version of this figure carved of granite in his exhibition 'After Cosmati' at the Royal Academy.

Method:

1. Make a dodecahedron (see page 38) – or use one previously made.

2. Make TWELVE pentagonal pyramids, the base of which follows the dotted line in Figure 6.11 above – the outer pentagon being the faces of the dodecahedron. The net for the pyramids is shown in Figure 6.12 above.

3. Assemble the twelve pyramids. Stick them on to the twelve faces of the dodecahedron to create the compound of the Dodecahedron plus Icosahedron.

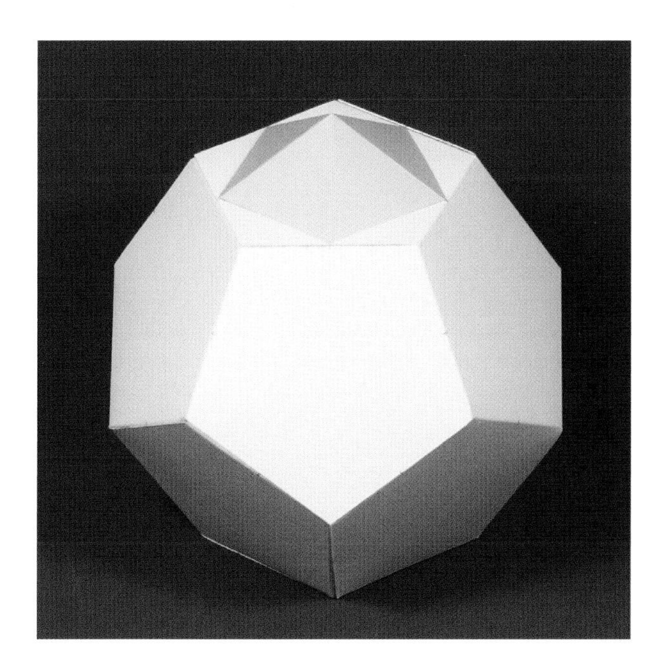

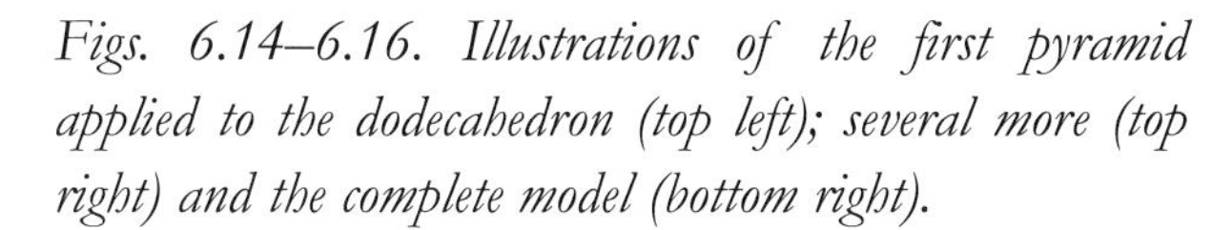

Figs. 6.14–6.16. Illustrations of the first pyramid applied to the dodecahedron (top left); several more (top right) and the complete model (bottom right).

Euclid's Method for Making a Dodecahedron

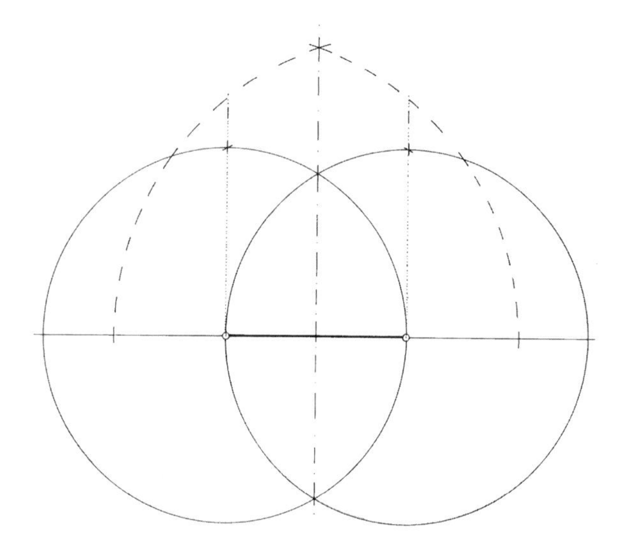

Fig. 6.17. First step in constructing a pentagon on the line.

Method:

Involves adding a 'roof' to each side of a cube.

1. First make a cube (see page 36). Then, taking the side length of the cube as the base line, construct a pentagon as shown in Figures 6.17 and 6.18 (see also page 67). Figure 6.19 shows the isolated pentagon, with the internal pentagram drawn, as well as certain diagonals of the innermost pentagon (two of which are extended to the edge of the larger pentagon).

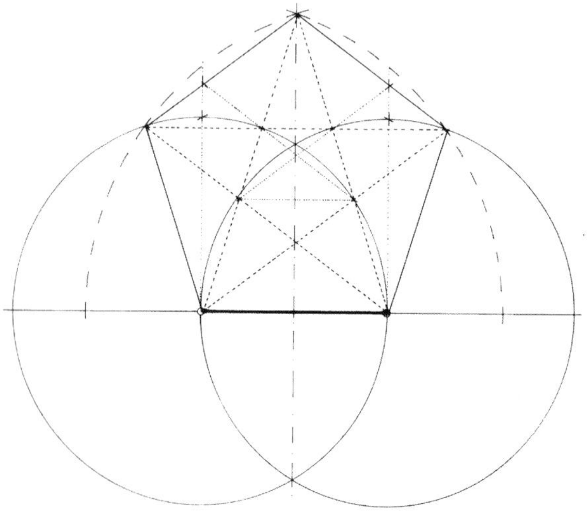

Fig. 6.18. The completed pentagon.

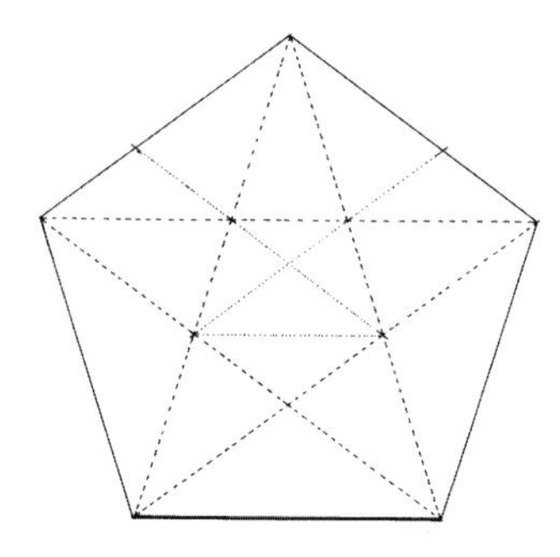

Fig. 6.19. The pentagon with construction lines removed.

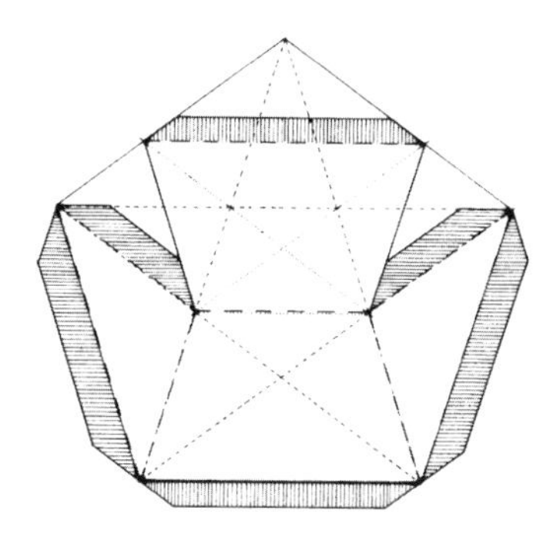

Fig. 6.20. The net for each roof.

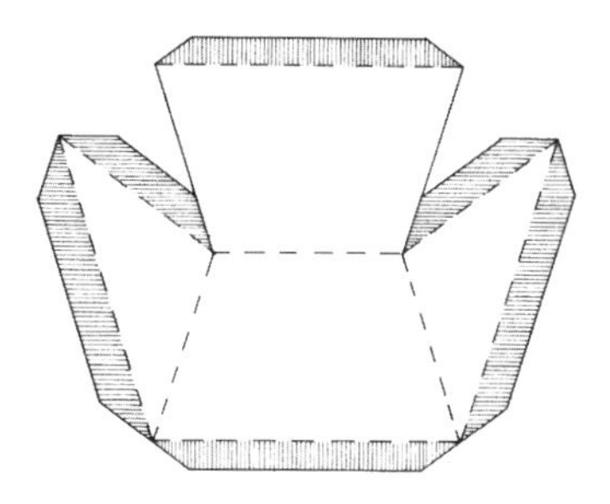

Fig. 6.21. The net for each roof with construction lines removed.

2. This gives all the lines and points necessary from which to construct the net for each roof. Figure 6.20 shows how this is done, together with the addition of tabs. Figure 6.21 shows the net without any construction lines.
   You will need SIX of these.

3. Cut, score and fold the nets. When adding the roofs to the cube, first experiment with how the model will look. The scores need to be folded tight so that they do not push up from the cube, opening up the edge line. If you are using different colours for the cube and the roofs, you might find it works better to have some tabs folded in, and some folded out (over the edge of the cube) so that the cube colour does not show through.

Fig. 6.22 The completed model.

Small Stellated Dodecahedron

The first stellation of the Dodecahedron (see Appendix A for an explanation of 'stellation').

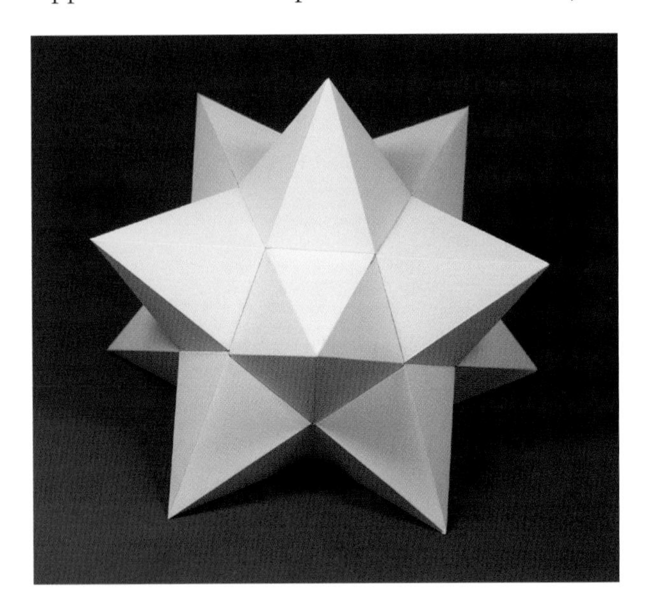

Method:

Involves adding a 'stellation' to each face of a Dodecahedron.

1. First make a dodecahedron (see page 38), or use one you have previously constructed.

2. Then, taking the edge length of the dodecahedron as the length AB (see Figure 6.23), construct a pentagon as follows: draw horizontal through AB; then circles radius AB centred on first A then B. Draw the vertical through the intersection of the two circles: this gives midpoint X of AB. Construct perpendiculars from A and B to intercept the circles at D and C: join D and C to create square ADCB. From X draw a semicircle through D and C down to horizon at E and F. From A with radius AF draw an arcs up from F: then similarly with the same radius from B. The arcs cross on the vertical axis at O: and give you points G and H on the original circles. OHBAG is our pentagon.

3. From the apex O of your pentagon draw a circle with radius OA. Project the upper sides of the pentagon OG and OH down to meet the circle at J and K: and also upwards to meet the large circle at L and M. Bisect LOJ and MOK to find the diameter and final points of the net P and Q. In addition to the golden triangle AOB, you now have four similar triangles POJ, JOA, BOK and KOQ.

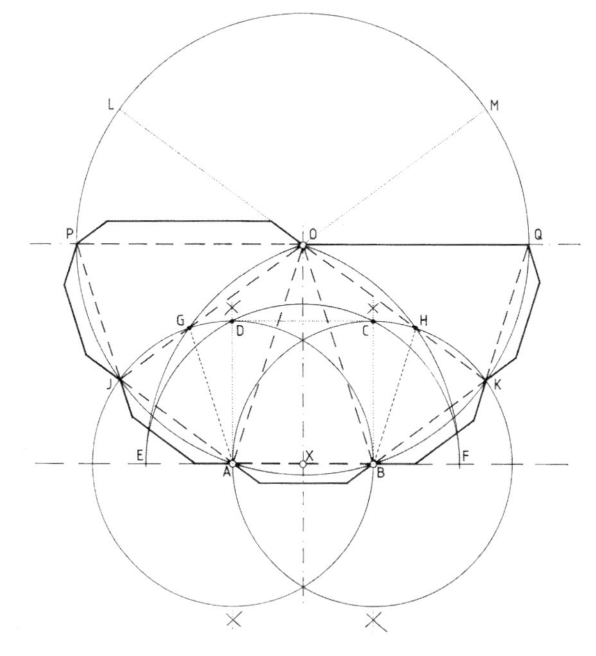

Fig. 6.23. Construction of the net for each stellation.

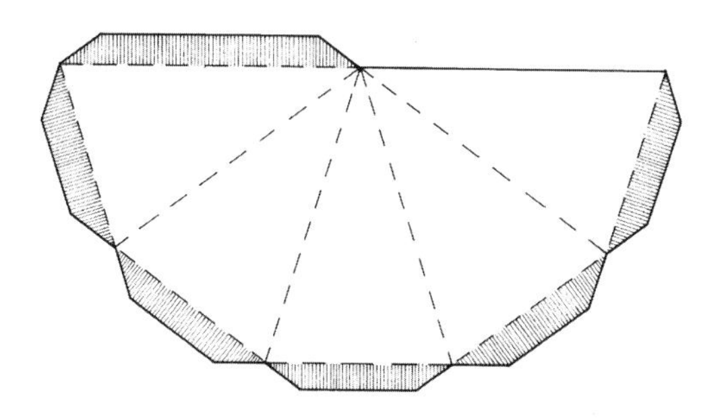

Fig. 6.24. The net for each stellation with construction lines omitted.

Note:

The same drawing can, with a slight modification, be used to yield two stellations (see Figure 6.25): but two loose tabs (as Fig. 6.26) will be needed to glue closed each stellation.

4. PQKBAJ is the net for one stellation.

5. Add the tabs and mark the cut and fold lines as shown – (see Figure 6.24 cleared of construction lines for clarity).

   You will need TWELVE of these.

6. Cut out along the solid lines: score along the dashed lines.

7. Fold each net, and glue together using the long tab (tab inside).

8. Glue one stellation to each face of the dodecahedron. Take your time so that each stellation has time to adhere fully, and is not dislodged when you stick on later stellations.

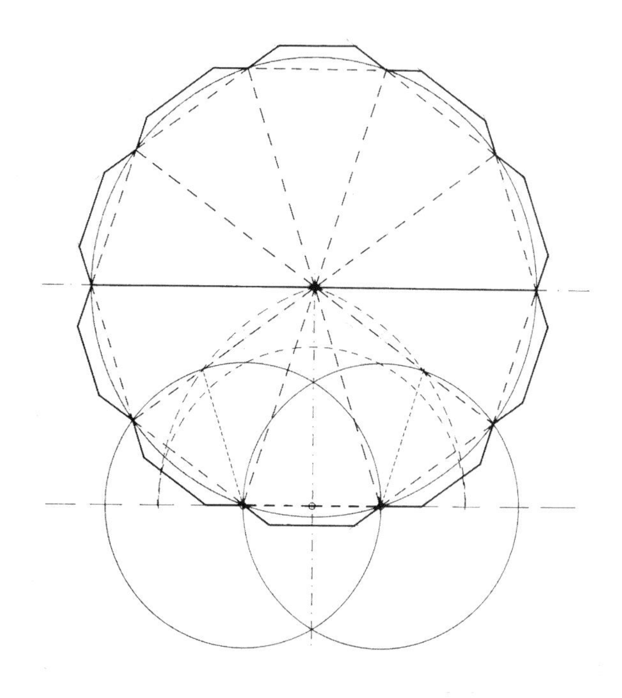

Fig. 6.25. Construction drawing to yield two stellations.

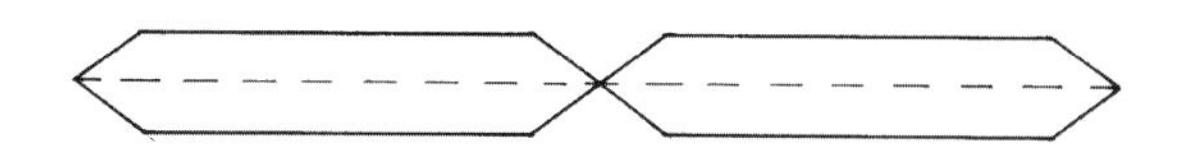

Fig. 6.26. The loose tabs required for each stellation.

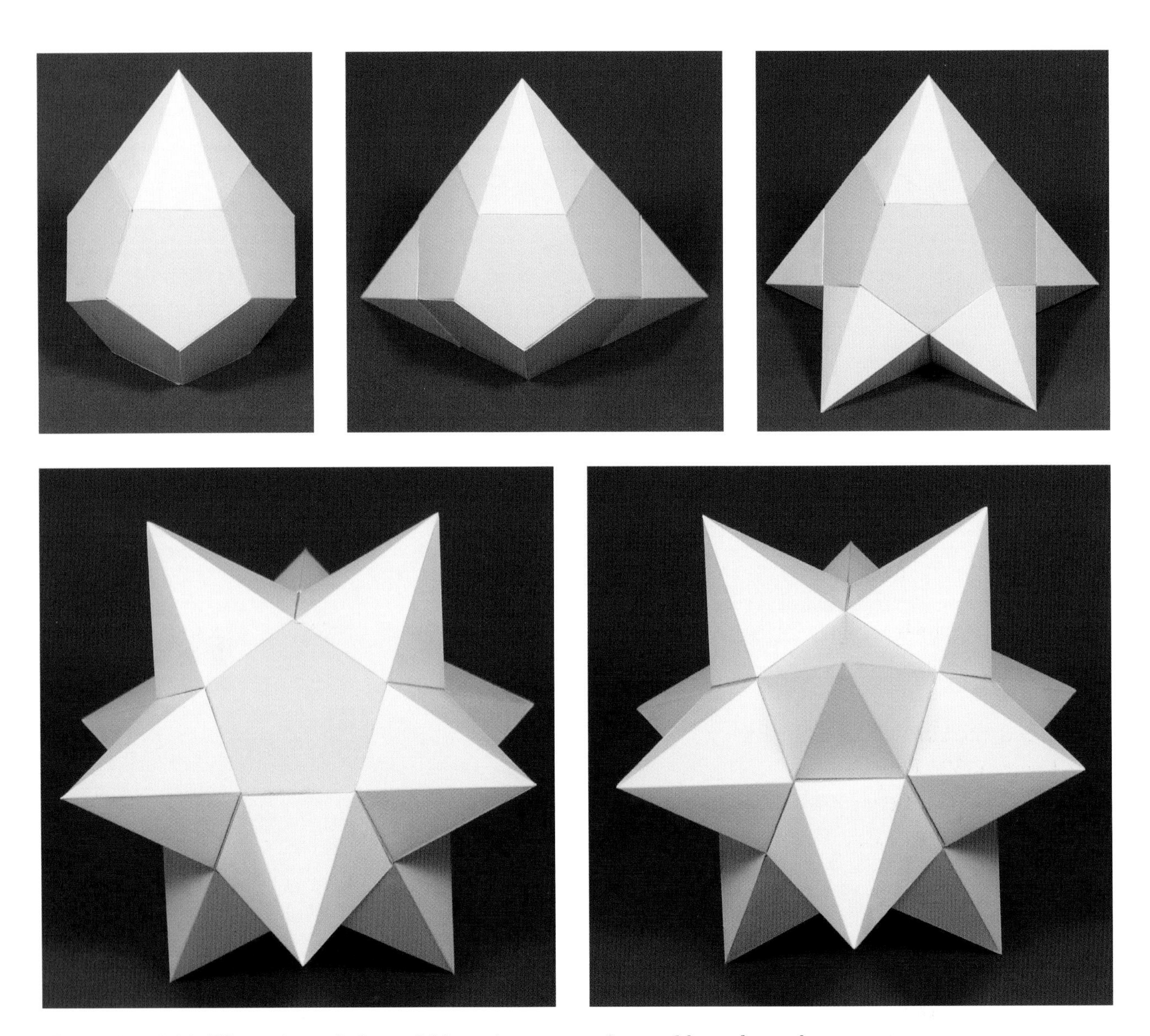

Figs. 6.27–6.31. Illustrations of the model in various stages of assembly, and complete.

Great Dodecahedron

The second stellation of the Dodecahedron, and one of the two Poinsot Polyhedra – the other being the Great Icosahedron (not featured here).

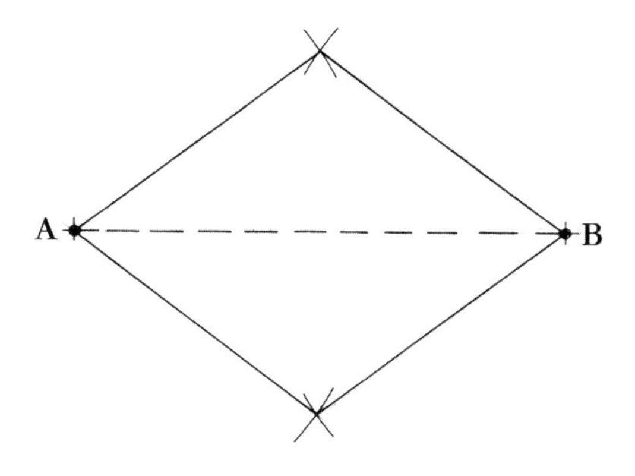

Fig. 6.32.

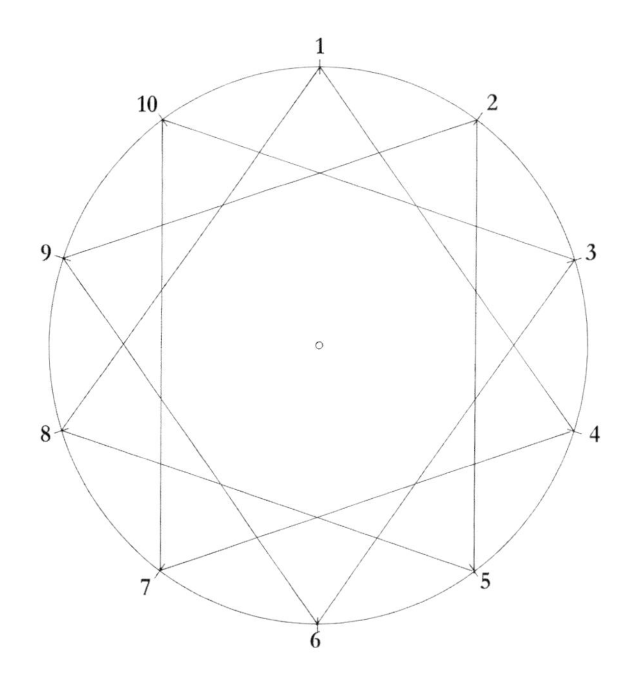

Fig. 6.33. Step 1.

Method:

1. This model can be made using the previously constructed Small Stellated Dodecahedron, and filling in each trough between apices with a folded rhomb – the net for which is given in Figure 6.32. This can be drawn by measuring the distance between apices plus the edge length of the stellations. Draw a line AB which is the distance between apices. Then with your compass on A, and radius the stellation edge length, draw arcs above and below the line. Do the same from B. The intersections of the arcs give you the required rhomb – (which you will find is a Golden Rhomb: that is, the diagonals are in the ratio 1: Φ).

   You will need TWENTY of these.

2. Alternatively, you can use the following construction to draw the net of five Golden Rhombs:

3. (Step 1) First construct a decagon – (see page 74). Then join the points of the decagon by missing two – that is, in the sequence 1, 4, 7, 10, 3, 6, 9, 2, 5, 8, 1 (Fig. 6.33).

4. (Step 2) By using the star lines drawn, construct the net for the five rhombs as shown in Figure 6.34. The solid lines are cut lines; the

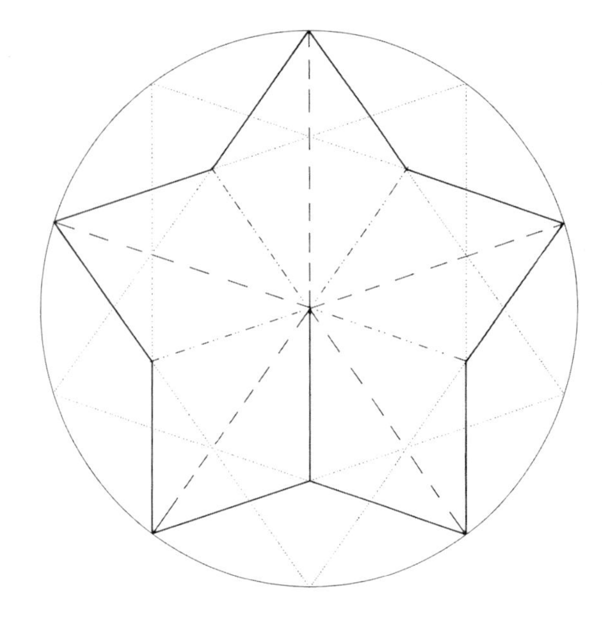

Fig. 6.34. Step 2. Don't forget the cut line from the centre to the edge of the net. Without this cut it will be difficult to fold.

dashed lines are to be folded one way (mountain folds); the dashed/dotted lines are to be folded the other way (valley folds). The resulting shape when cut, scored and folded can be seen in Figure 6.35 below.

You will need FOUR of these.

5. Cut out along the solid lines: score along the fold lines. Fold each net to create a folded star.

6. Glue the folded stars to the Small Stellated Dodecahedron one by one. You might find it easier to glue the central spike first, flapping the two spikes on each side of it away. When this has taken, glue the two spikes on one side, and then the other. The Great Dodecahedron will slowly emerge.

Figs. 6.35–6.36. Illustrations of the folded star net of five folded rhombs (above left); and with the first folded star glued to the core solid.

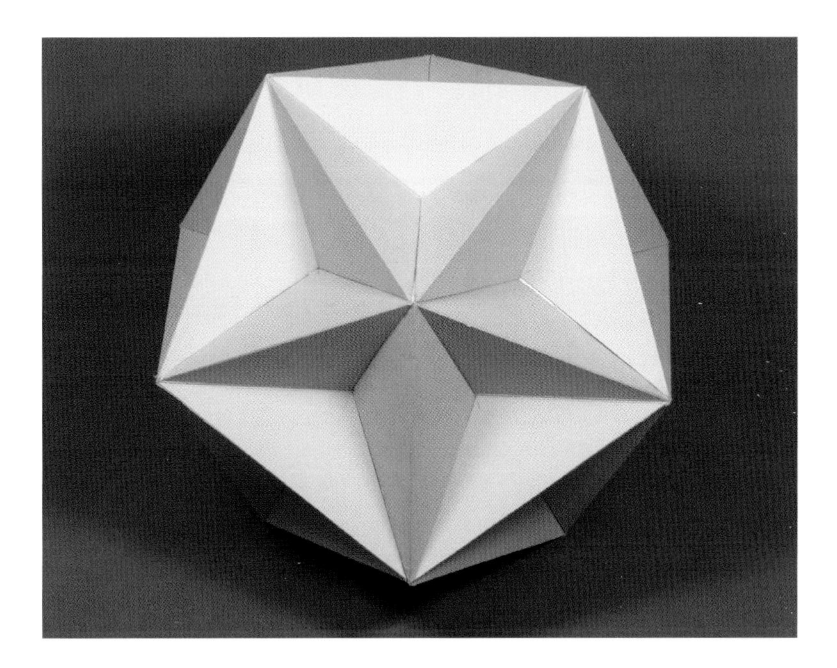

Figs. 6.37–6.38. Two very different views of the completed Great Dodecahedron taken vertex-on (top) and face-on (bottom). Note how the ridges of the stellations form an Icosahedron.

Great Stellated Dodecahedron

The third stellation of the Dodecahedron.

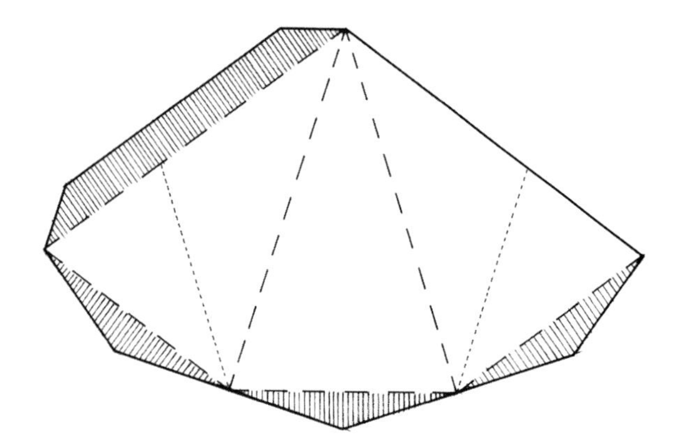

Fig. 6.39. The net for each stellation.

Method:

1. Using the same construction as for the Small Stellated Dodecahedron (see page 102) – but using just three of the golden triangles for each stellation (see Figure 6.39), though with steeper tab angles to avoid overlapping when folded inside – the Great Stellated Dodecahedron can be made by adding one such stellation to each of the twenty faces of an icosahedron. You can see why the icosahedron is used as the core by looking at the previous construction, which resulted in an icosahedral form (see Figure 6.38 on previous page). The same result would be obtained if one used a Great Dodecahedron as the core on to which to plant the stellations – but the icosahedron gives a better platform.

   You will need TWENTY of these.

2. To help in the production of the twenty stellation nets required, you can use the construction on page 102, but double it, to produce a drawing that will yield six nets each time – one

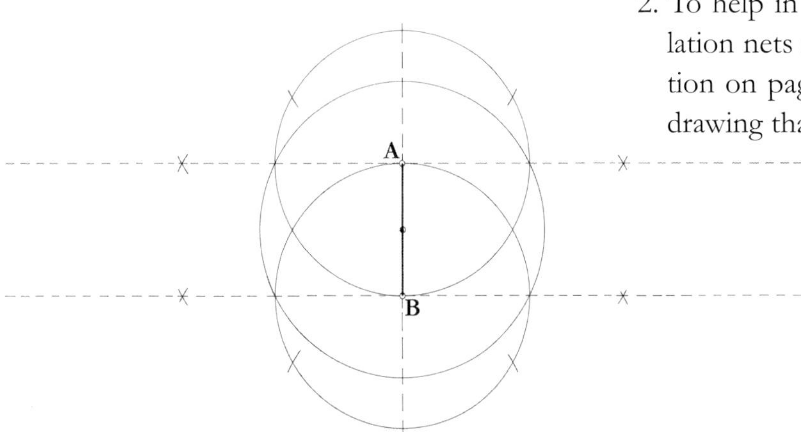

Fig. 6.40. Step 1. Begin with the known edge length AB – (which is the edge length of the base icosahedron) – and refer to the construction on page 102 – except that we will be drawing a double construction (the lefthand side mirrored on the righthand side). To do this, the construction is best turned through 90 degrees.

double and four singles (see the construction drawing sequence in Figures 6.40 to 6.44). The singles will require loose tabs to make up the pyramids. THREE of these double decagon nets, plus TWO further single nets, will be needed.

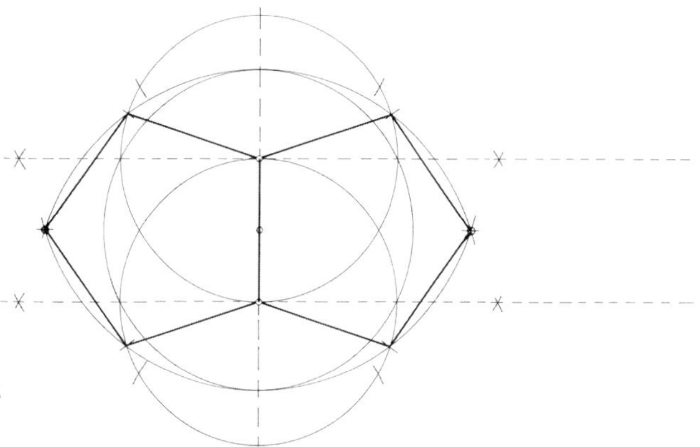

Fig. 6.41. Step 2. Two back to back pentagons appear.

3. When you have drawn out the nets for all twenty stellations, cut along the solid lines: score along the dashed lines. Fold each net, and glue together using the long tab or a loose tab (tabs inside).

Note: in the double nets the centre score line where the two decagons meet will be folded the other way (as a valley fold).

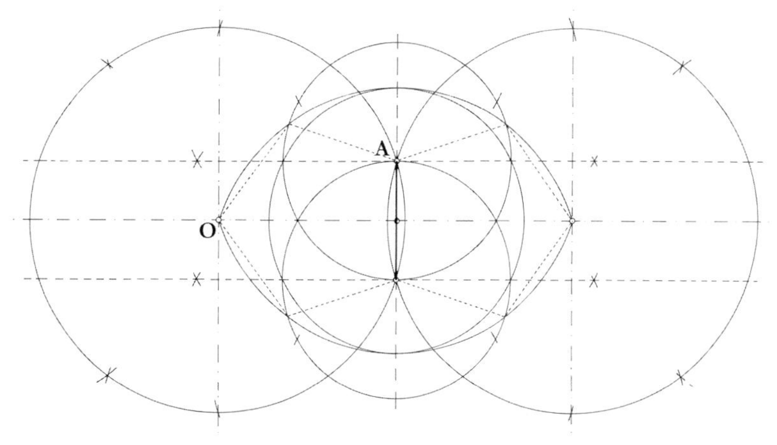

Fig. 6.42. Step 3. The apices of the pentagons are the centres of circles, radius OA, circumscribing the required decagons. The edge length can then be stepped round the circle to find all the points of the decagons on the circles.

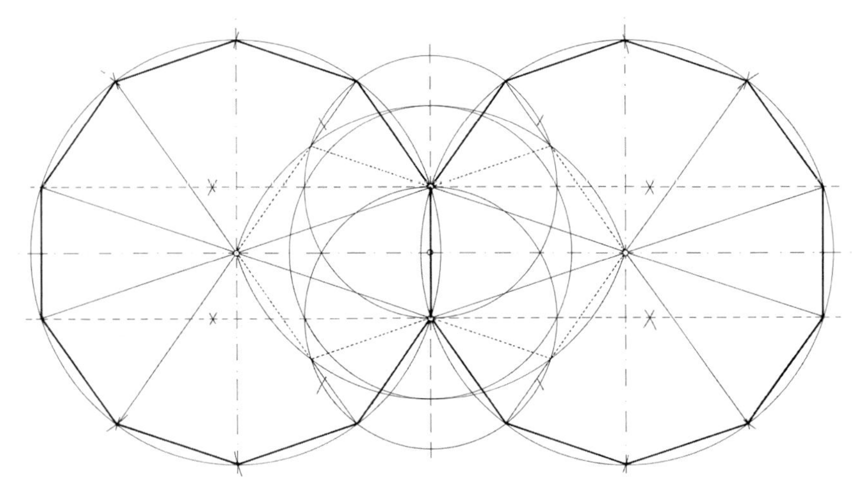

Fig. 6.43. Step 4. With the points of the two decagons now all found, the radial cut/fold lines can be easily added to produce the multiple nets below.

4. Progressively, glue one stellation to each face of the icosahedron. Take your time to allow the glue to take hold.

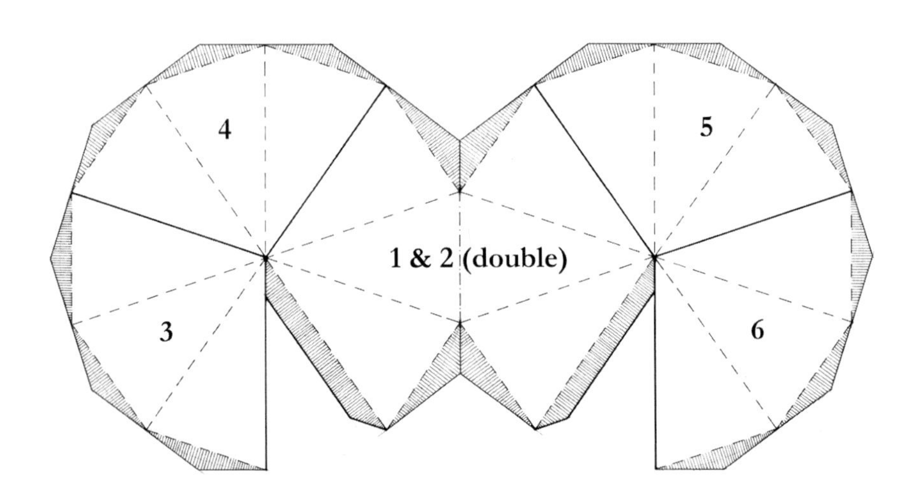

Fig. 6.44. On this drawing all construction lines have been removed for clarity, showing only the nets of the six stellations. Note: the nets numbered 3, 4, 5 and 6 will each need a loose tab.

Figs. 6.45–6.49. Illustrations of (top left) the core icosahedron with one double and two single pyramids made up; the model in various stages of assembly; and (bottom right) complete.

Net for the Rhombic Dodecahedron

12 (√2 rhomb) faces, 14 vertices, 24 edges. Dual of the Cuboctahedron.
This and the following solid are known as the Kepler Solids.

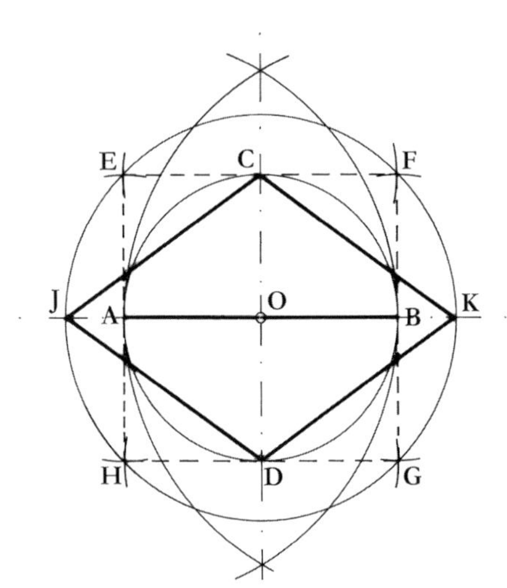

Fig. 6.50. Step 1

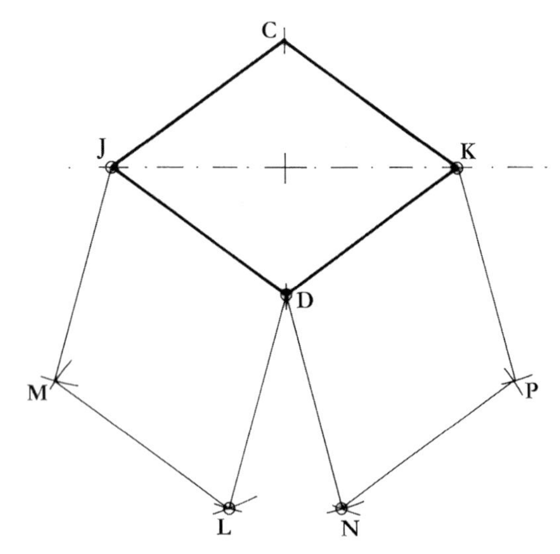

Fig. 6.51. Step 2.

Method:

1. (Step 1) To fit the whole drawing on the page, draw the first part of the construction (as shown in Fig. 6.50) in the lefthand third of your sheet. Draw a horizontal line across the page just above halfway, and mark AB on this. Find the midpoint O by drawing arcs of radius AB from A and B in turn, and joining the points of intersection. Draw a circle centre O radius OA. Construct a square about this circle – (the corners are found by drawing arcs of radius OA from A, C, B and D). Then draw a circle centre O radius OE. This crosses the horizontal at J and K. JCKD is the required rhomb, whose diagonals CD and JK are in the ration 1:√2 (being derived from the side and the diagonal of the square).

2. (Step 2) The rest of the construction is made by drawing a series of arcs using only two radii – JC and JK. If you have two pairs of compasses, it will help to use both, with each set to one of the required radii. Draw arcs of radius JC from D and radius JK from J to find point L. Then draw arcs radius JC from points J and L to find M. Similarly, JC from D and JK from K find N: then JC from K and N to find P. JDLM and DKPN are two further √2 rhombs (Fig. 6.51).

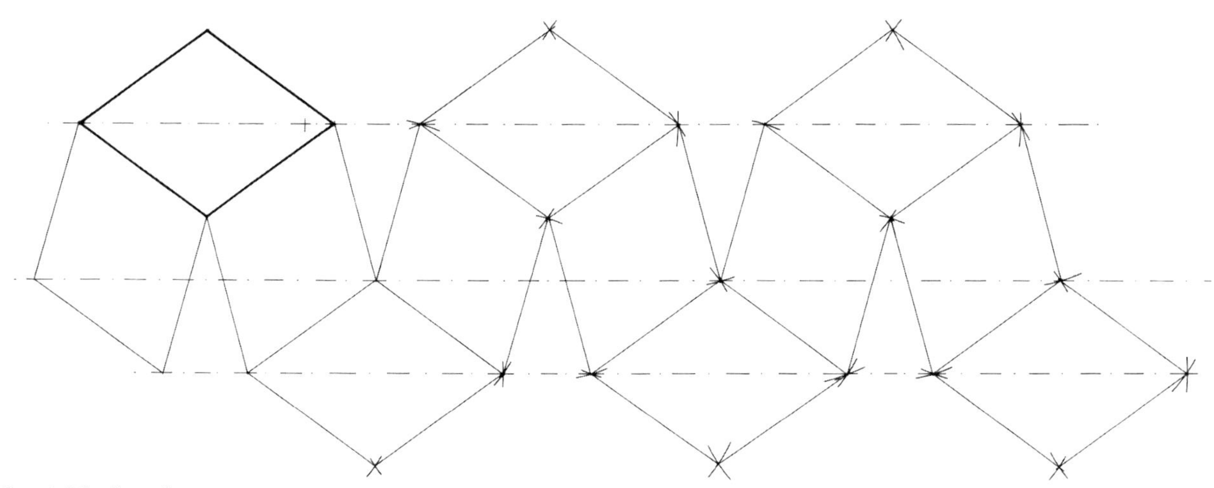

Fig. 6.52. Step 3.

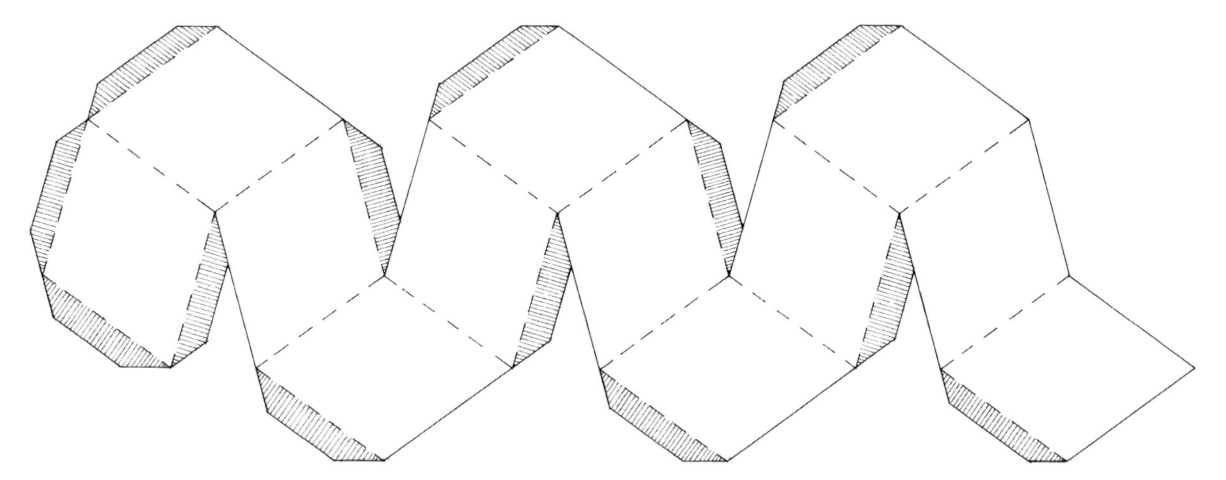

Fig. 6.53. Step 4.

3. (Step 3) Repeat the process across the page until you have the complete net (Fig. 6.52). You might find it helpful to draw horizontals across the page to fine tune the arc intersections and keep your drawing accurate.

5. (Step 4) Add tabs and mark the cut and fold lines as shown (Fig. 6.53).

6. Cut out along the solid lines: score along the dashed lines. Fold upwards along the scored lines. Glue carefully and progressively (tabs inside) – one or two tabs at a time. Clean up and decorate as you wish.

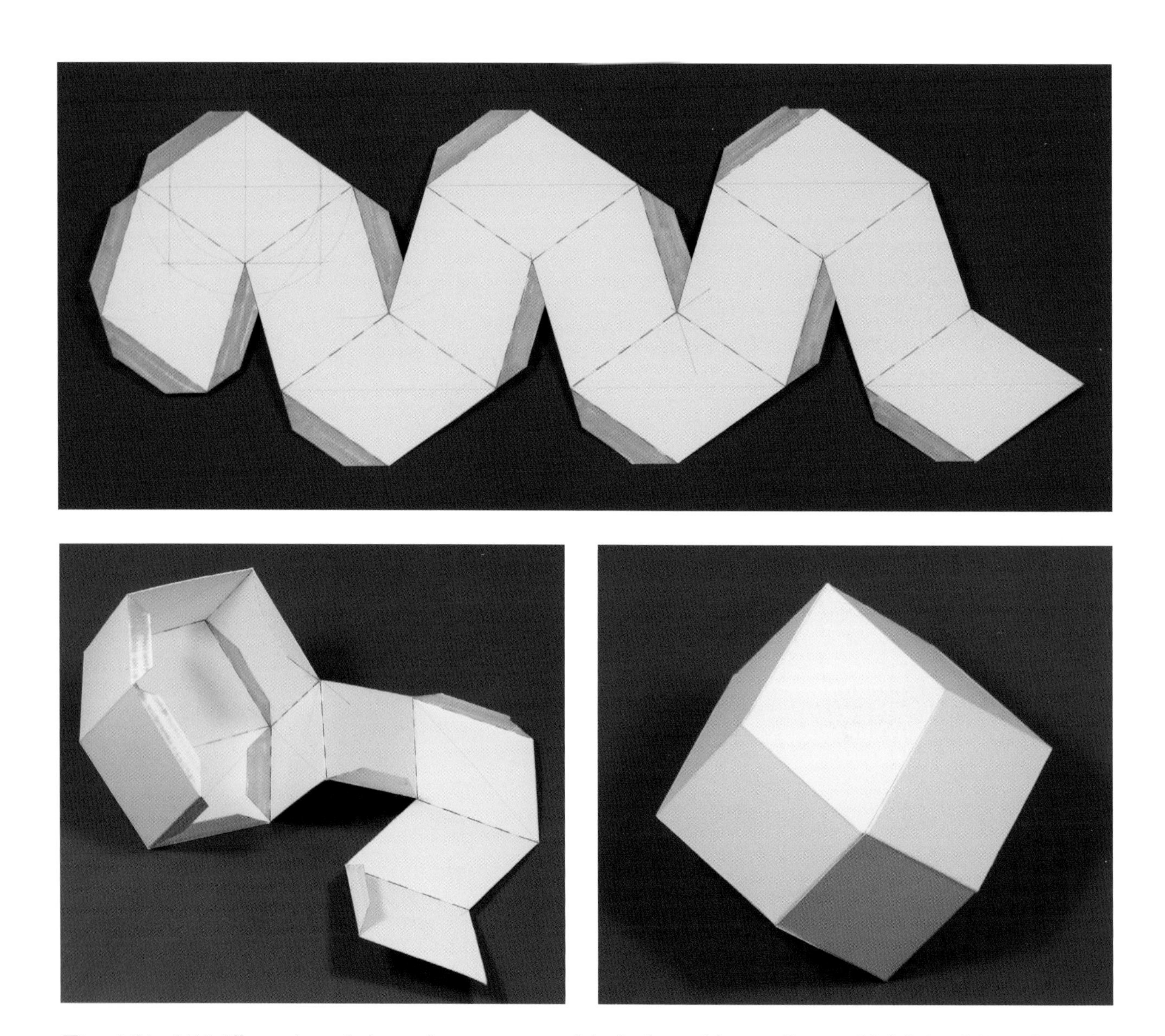

Figs. 6.54–6.56. Illustrations of the net drawn out on card (top); the model partially assembled (below left); and complete (bottom right).

Fig. 6.57. Illustration of the Cube/ Octahedron Compound sitting exactly within the Rhombic Dodecahedron (in green). The apices of the inner solid are the apices of the surrounding Rhombic Dodecahedron.

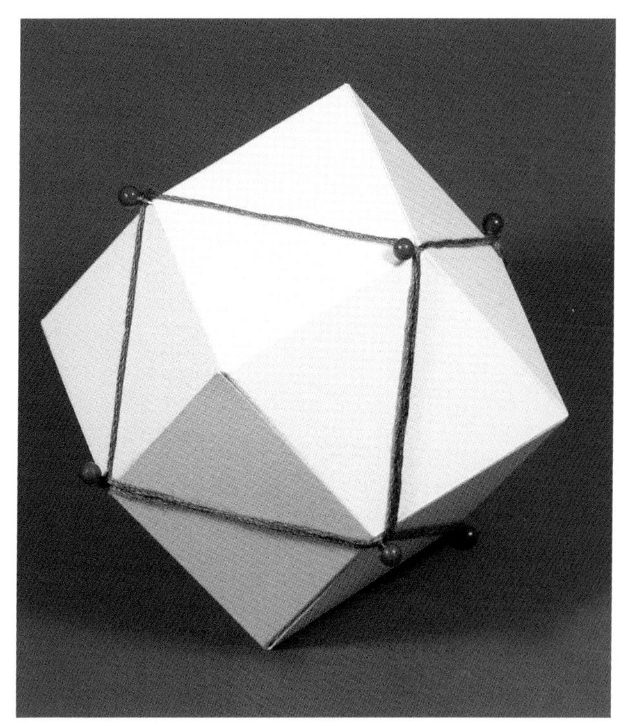

Fig. 6.58. If you trace the short diagonals around the solid, you find a Cube. Similarly, the long diagonals would give you an Octahedron.

Net for the Rhombic Triacontahedron

30 (Golden Rhomb) faces, 32 vertices, 60 edges. Dual of the Icosidodecahedron. This and the previous solid are known as the Kepler Solids.

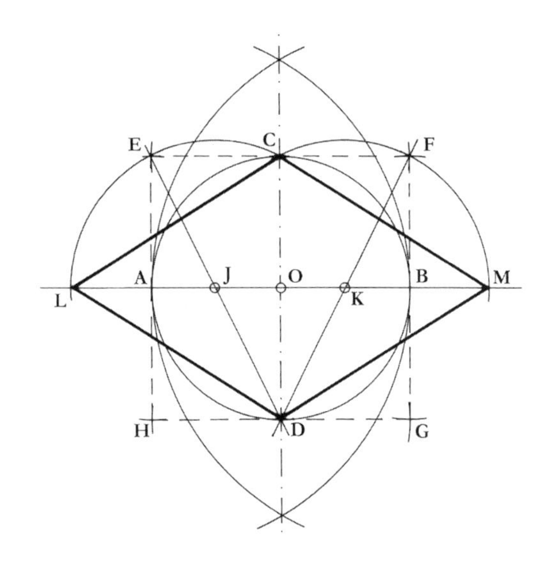

Fig. 6.59. Step 1.

Method:

1. The method is similar to that for the Rhombic Dodecahedron, except that a different rhomb is used – and the net is larger, so I suggest it is made in parts.

2. (Step 1) As before (see page 112) construct a circle (centre O the midpoint of AB) and then a square (EFGH). Draw DE and DF to find the midpoints of OA and OB. With centre J draw an arc radius JE down to the horizontal to find L. Likewise, an arc with centre K from F to M. LCMD is the required rhomb, whose diagonals CD and LM are in the ration 1:Φ – being the side and the diagonal of a pentagon (Fig. 6.59).

3. (Step 2) Using only the two radii LC and LM (and two pairs of compasses if you have them), build the drawing as before (see page 112 and Fig. 6.60). (Step 3) Then add tabs and mark the cut and fold lines (Fig. 6.61).

You will need FIVE of these.

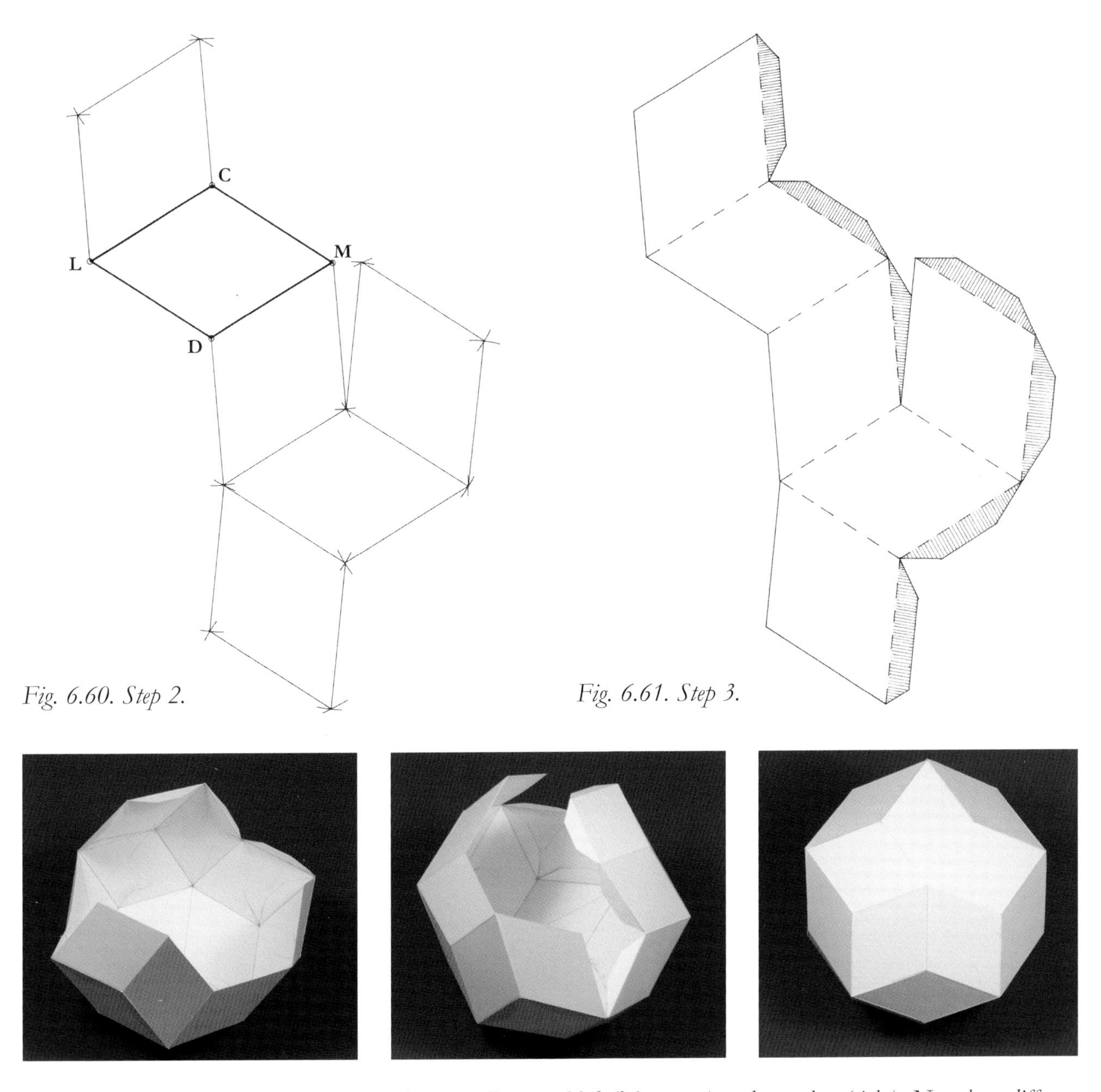

Fig. 6.60. Step 2.

Fig. 6.61. Step 3.

Figs. 6.62–6.64. Illustrations of the model partially assembled (left, centre) and complete (right). Note *how different appears this view compared to that at the top of the previous page, though they are pictures of the same model.* Note *also that if you trace the diagonals (as with the* Rhombic Dodecahedron*), the short diagonals give you a* Dodecahedron*, and the long an* Icosahedron*.*

Appendix A: Some Mathematical Terms

Compound

A regular polyhedral compound is a solid that combines several other regular polyhedra.

Dual

Every polyhedron can be paired with another polyhedron – its dual (also known as its reciprocal) – in which the vertices of one occur in the middle of the faces of the other.

Truncation

Truncation means slicing through each vertex in such a way that the sides of the resulting polygons are all equal (see Figure 7.1).

Snub

Snub means each face has been completely surrounded by equilateral triangles (see Figure 7.2).

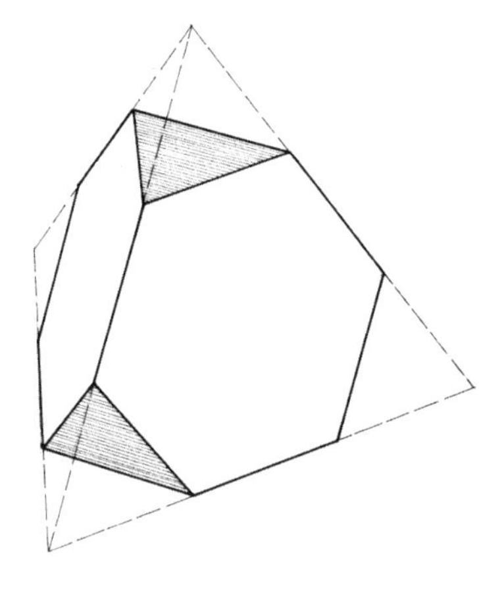

Fig. 7.1. Truncated Tetrahedron.

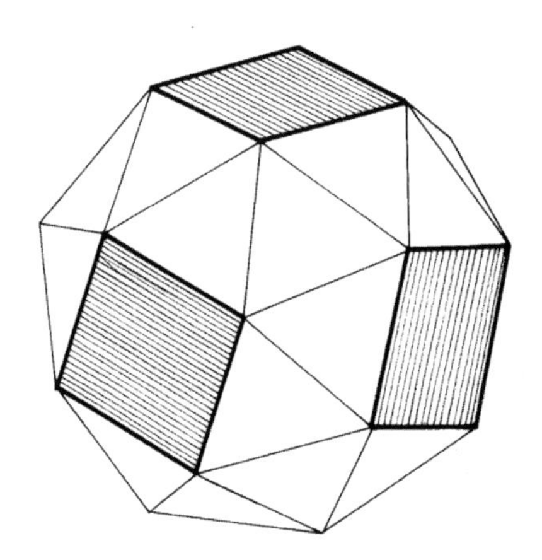

Fig. 7.2. Snub Cube.

Stellation

Stellation is a process of constructing new polygons in two dimensions, or new polyhedra in three dimensions, by projecting edges and folding up.

Symmetry

Symmetry can be visualised as follows:

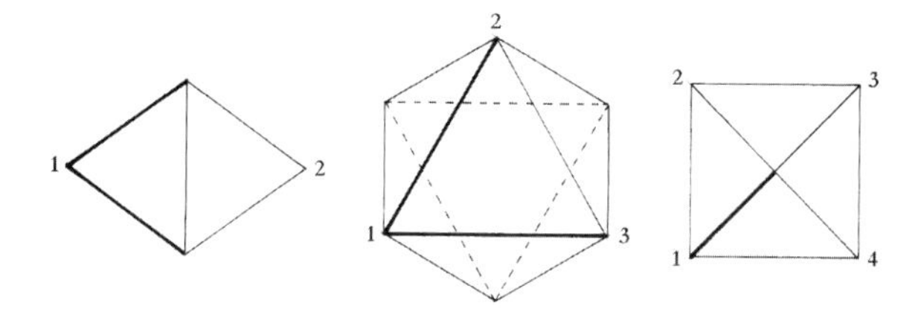

Fig. 7.3. Turn the edge, face and vertex-on views of the Octahedron above 2, 3 and 4 times in the diagrams to see its symmetries.

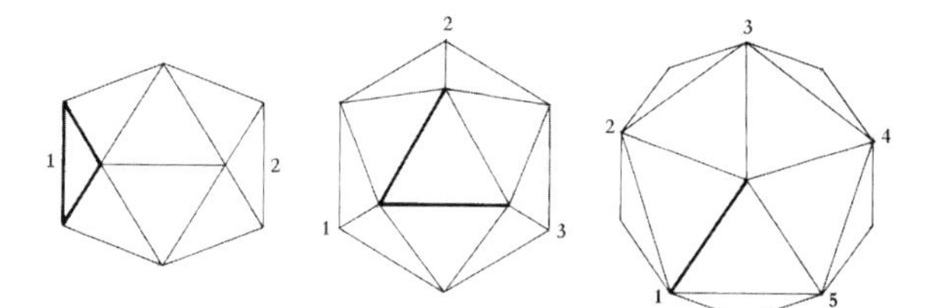

Fig. 7.4. Turn the edge, face and vertex-on views of the Icosahedron above 2, 3 and 5 times in the diagrams to see its symmetries.

Appendix B: Plain Nets for All Platonic and Archimedean Solids

Note: The solids can be 'opened up' to reveal a variety of patterns. Of the many, only one is given here for each solid.

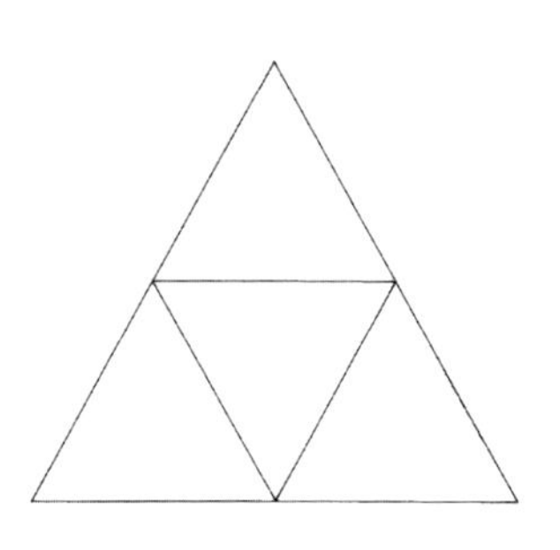

Fig. 8.1. Net of the Tetrahedron.

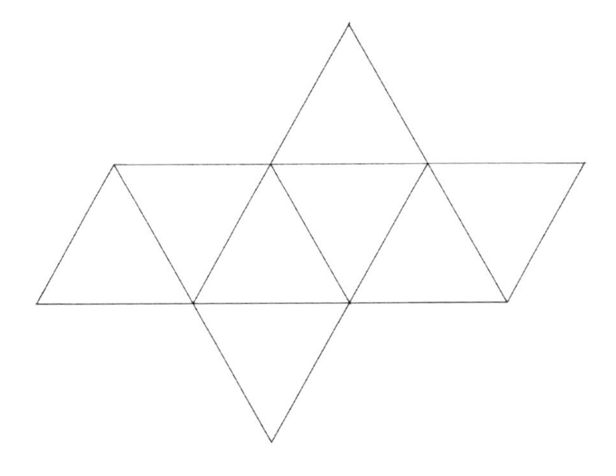

Fig. 8.2. Net of the Octahedron.

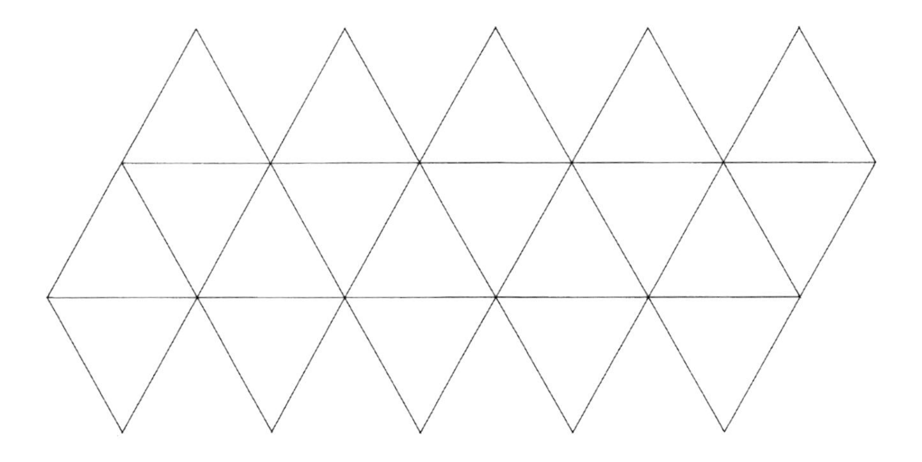

Fig. 8.3. Net of the Icosahedron.

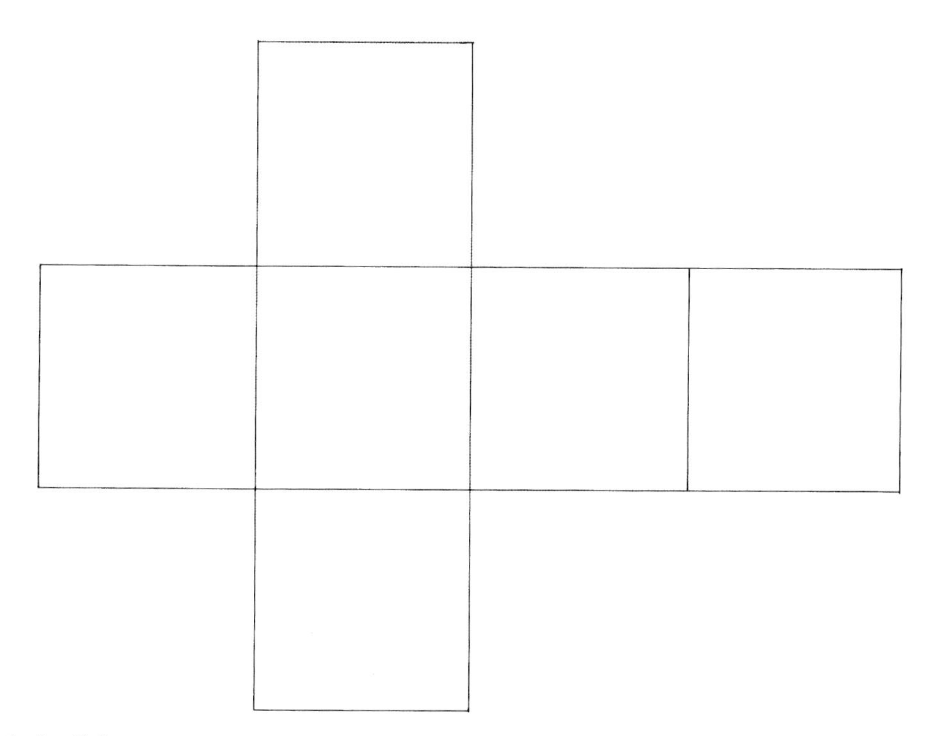

Fig. 8.4. Net of the Cube.

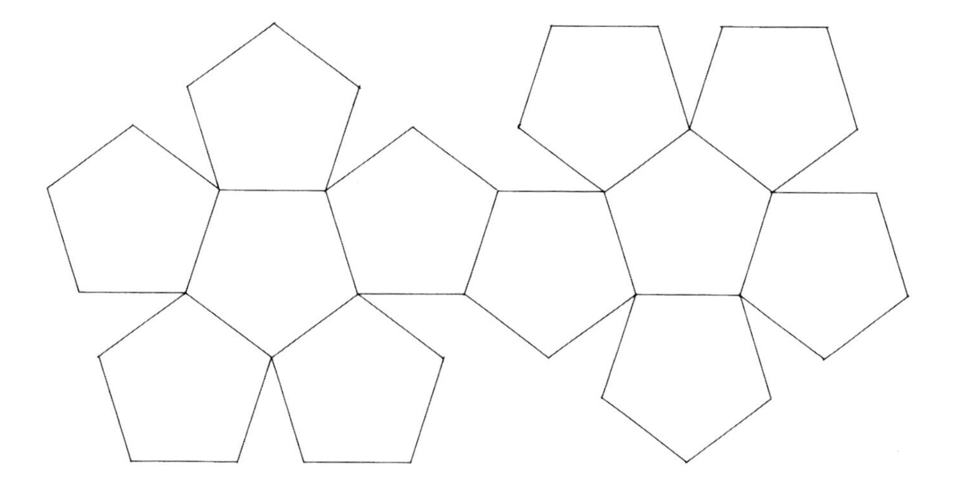

Fig. 8.5. Net of the Dodecahedron.

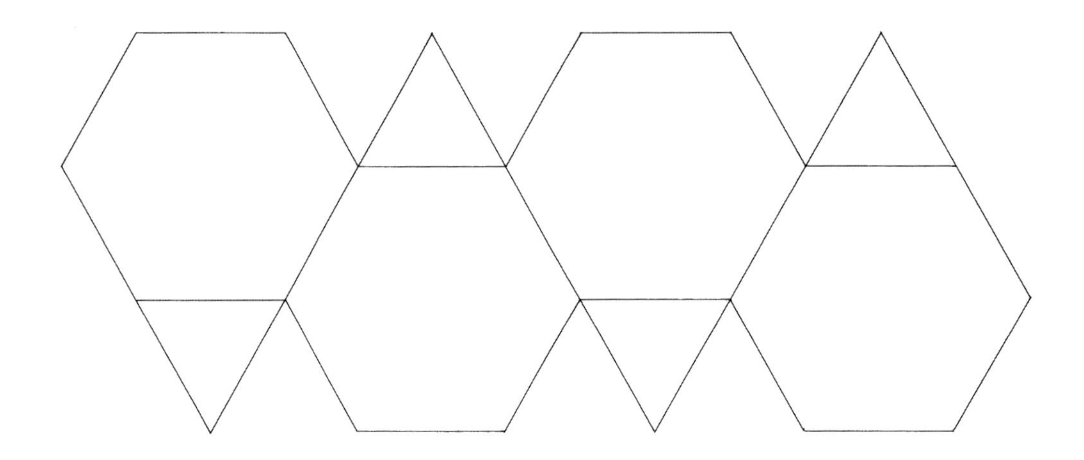

Fig. 8.6. Net of the Truncated Tetrahedron.

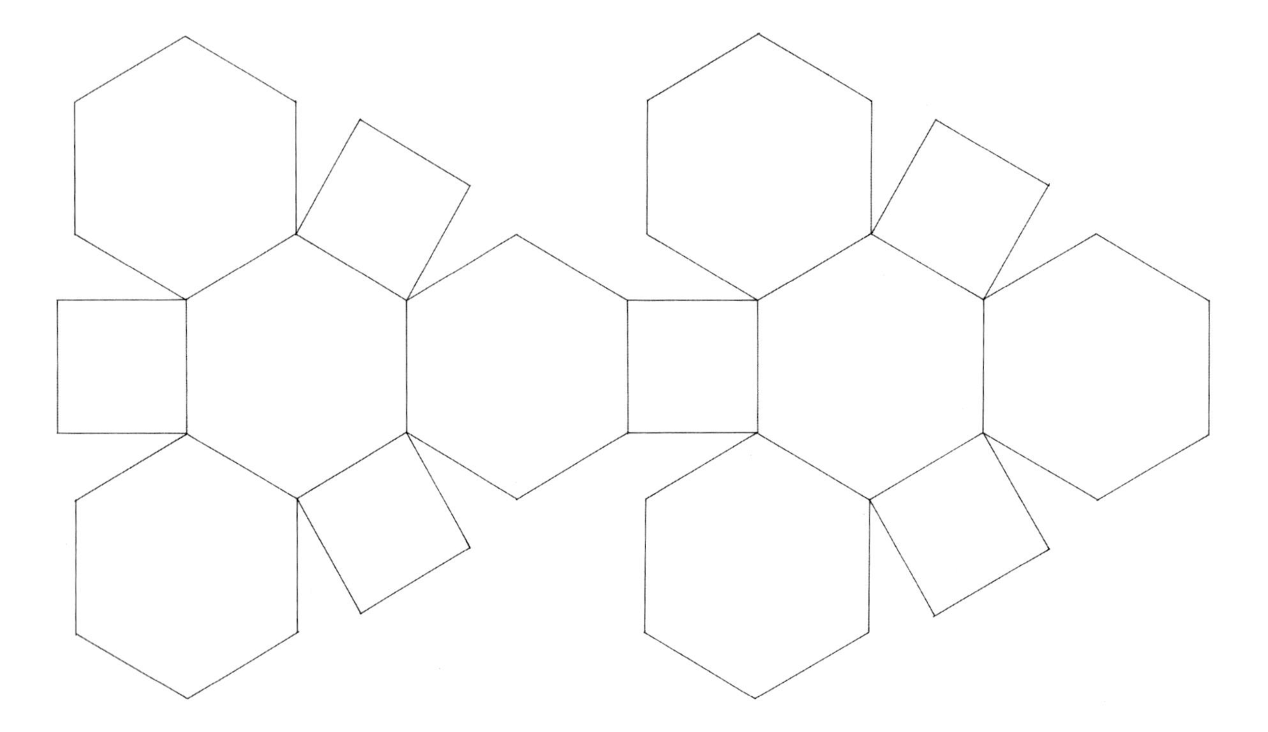

Fig. 8.7. Net of the Truncated Octahedron.

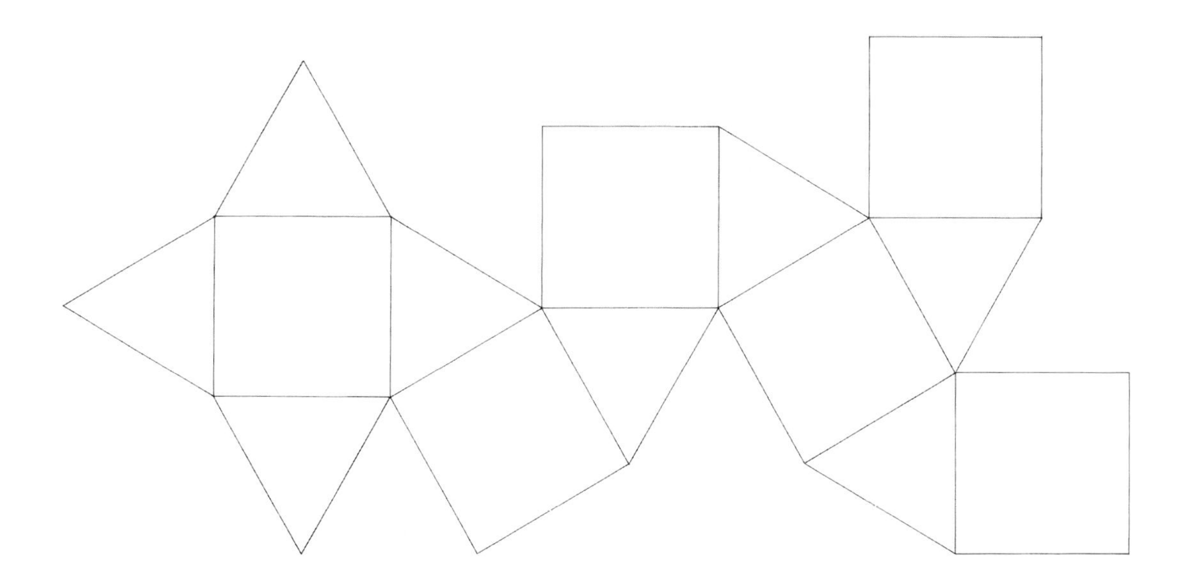

Fig. 8.8. Net of the Cuboctahedron.

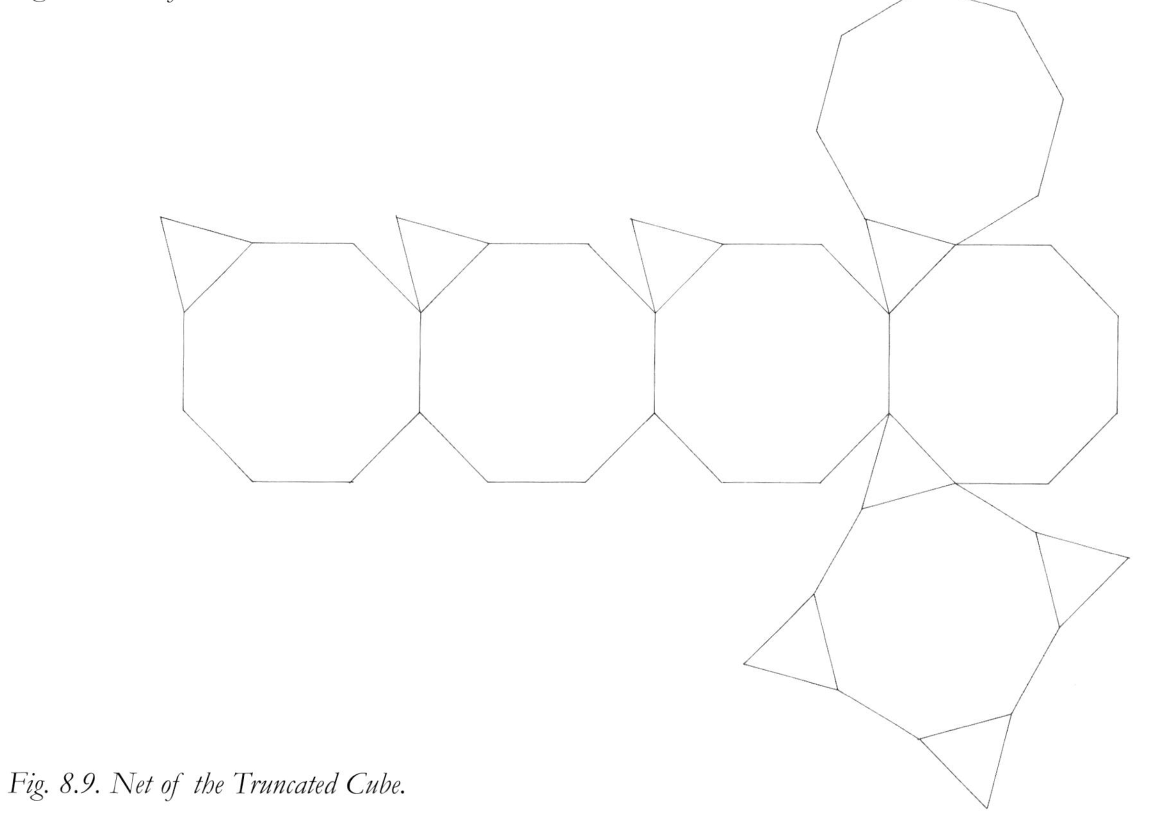

Fig. 8.9. Net of the Truncated Cube.

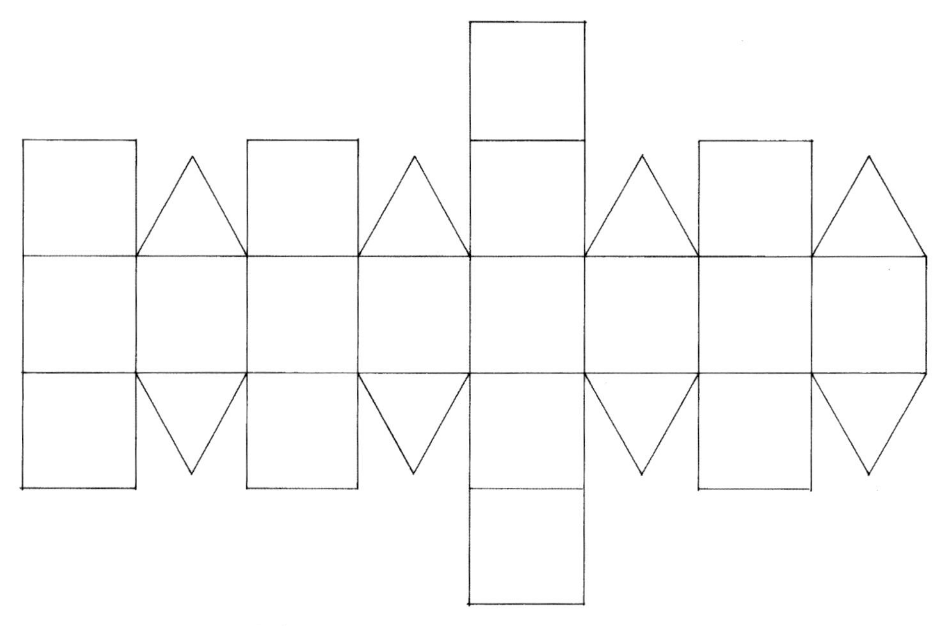

Fig. 8.10. Net of the Rhombicuboctahedron.

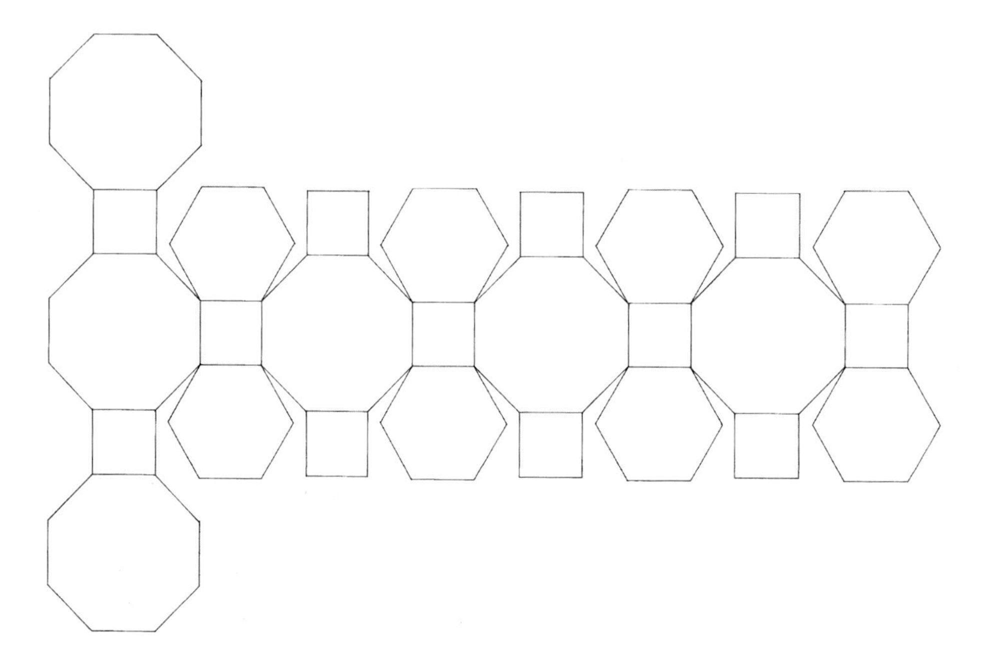

Fig. 8.11. Net of the Truncated Cuboctahedron.

Fig. 8.12. Net of the Icosidodecahedron.

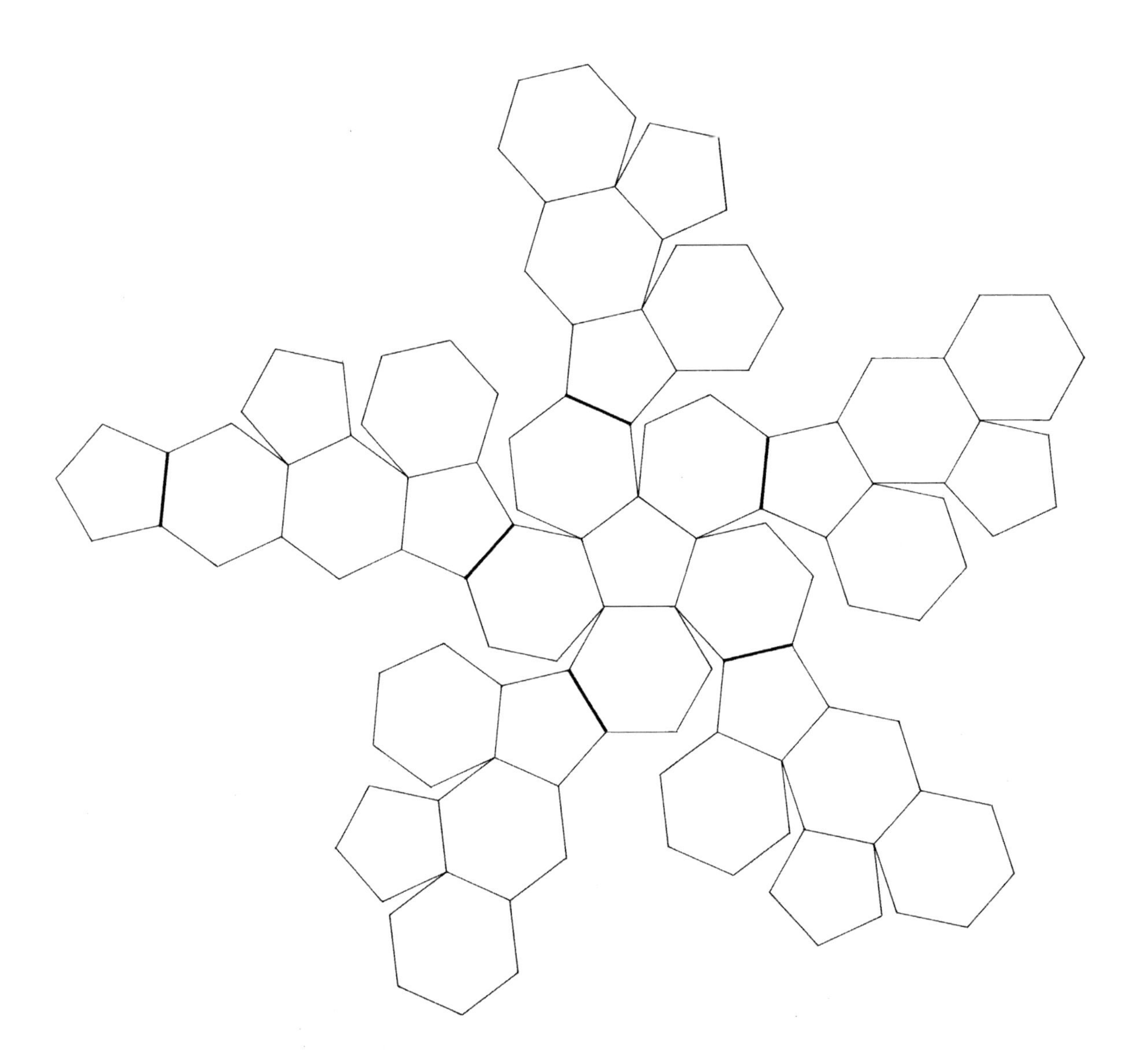

Fig. 8.13. Net of the Truncated Icosahedron.

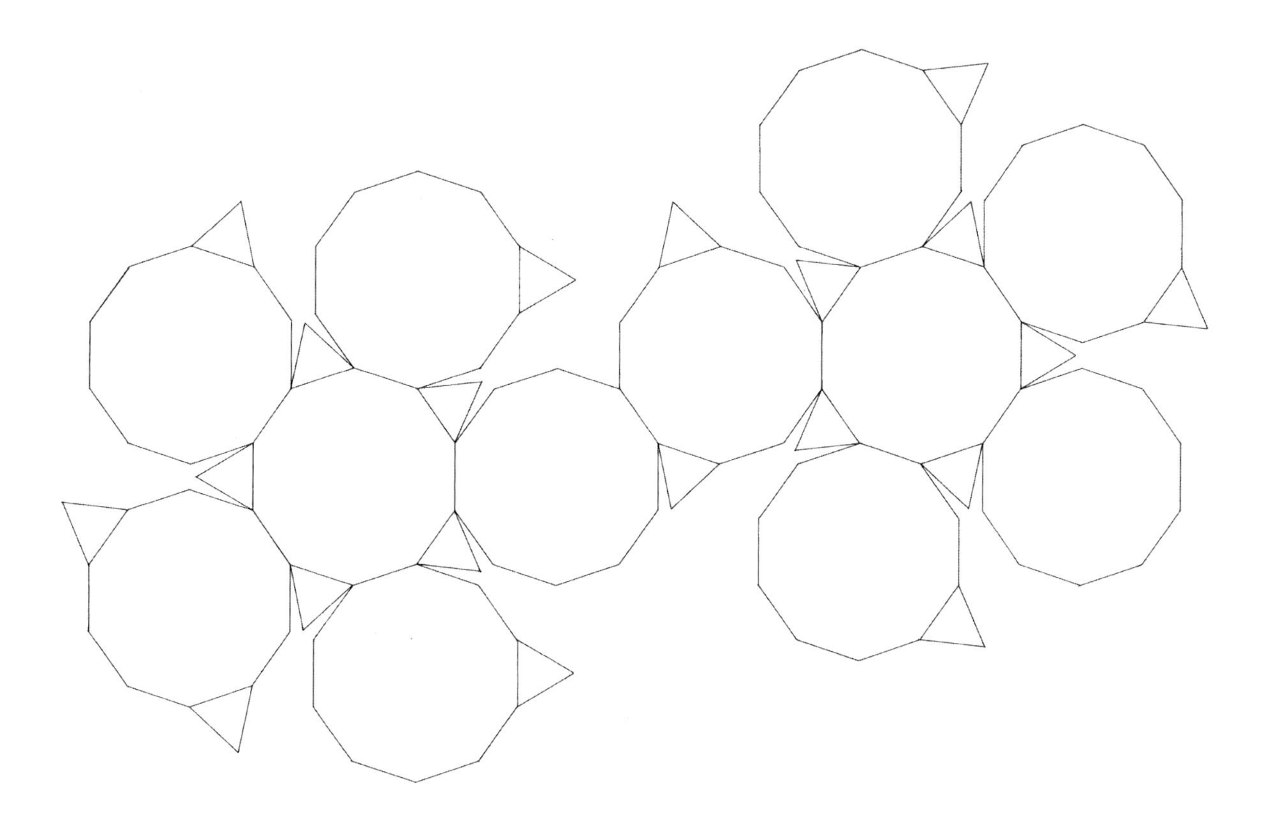

Fig. 8.14. Net of the Truncated Dodecahedron.

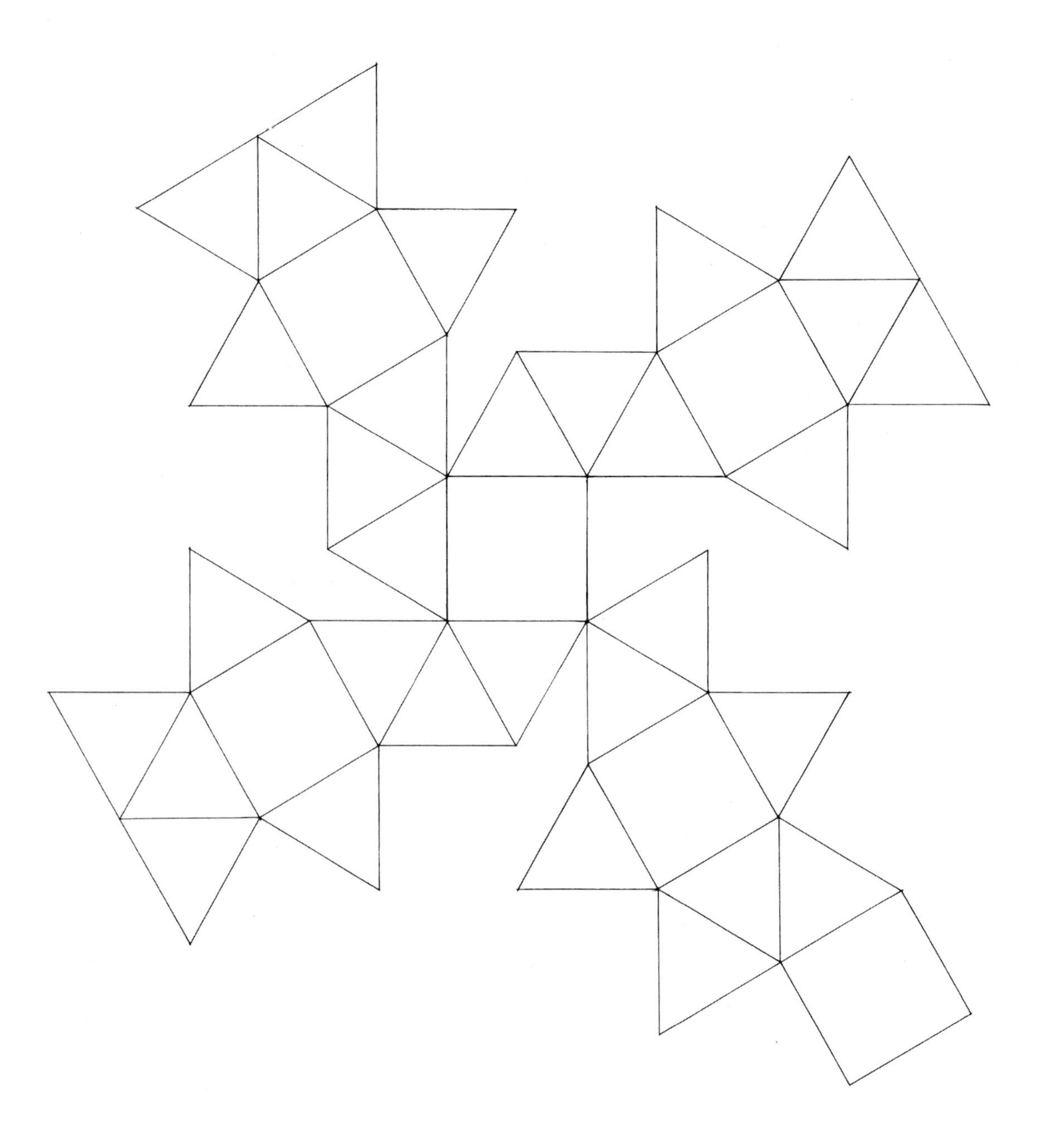

Fig. 8.15. Net of the Snub Cube.

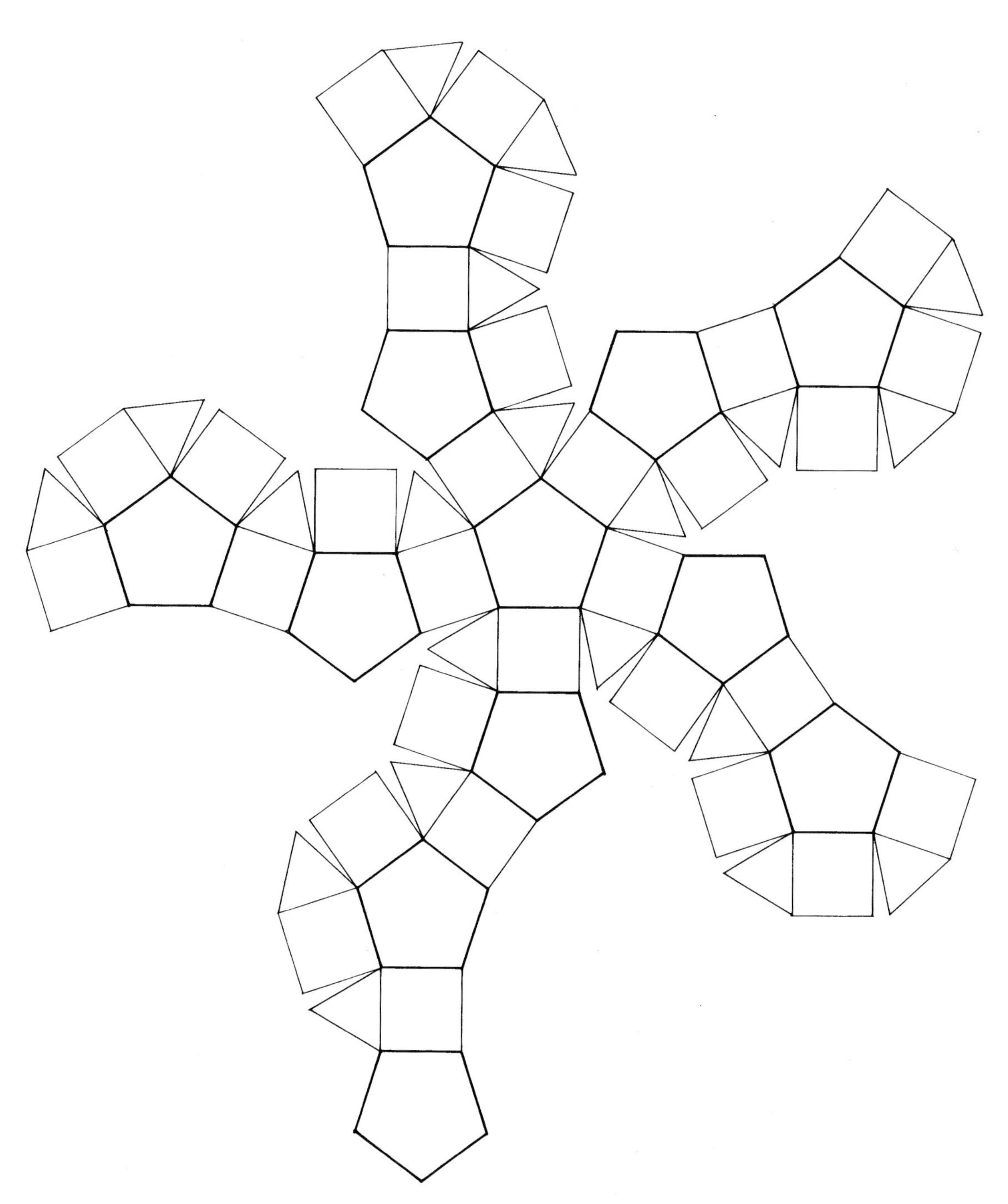

Fig. 8.16. Net of the Rhombicosidodecahedron.

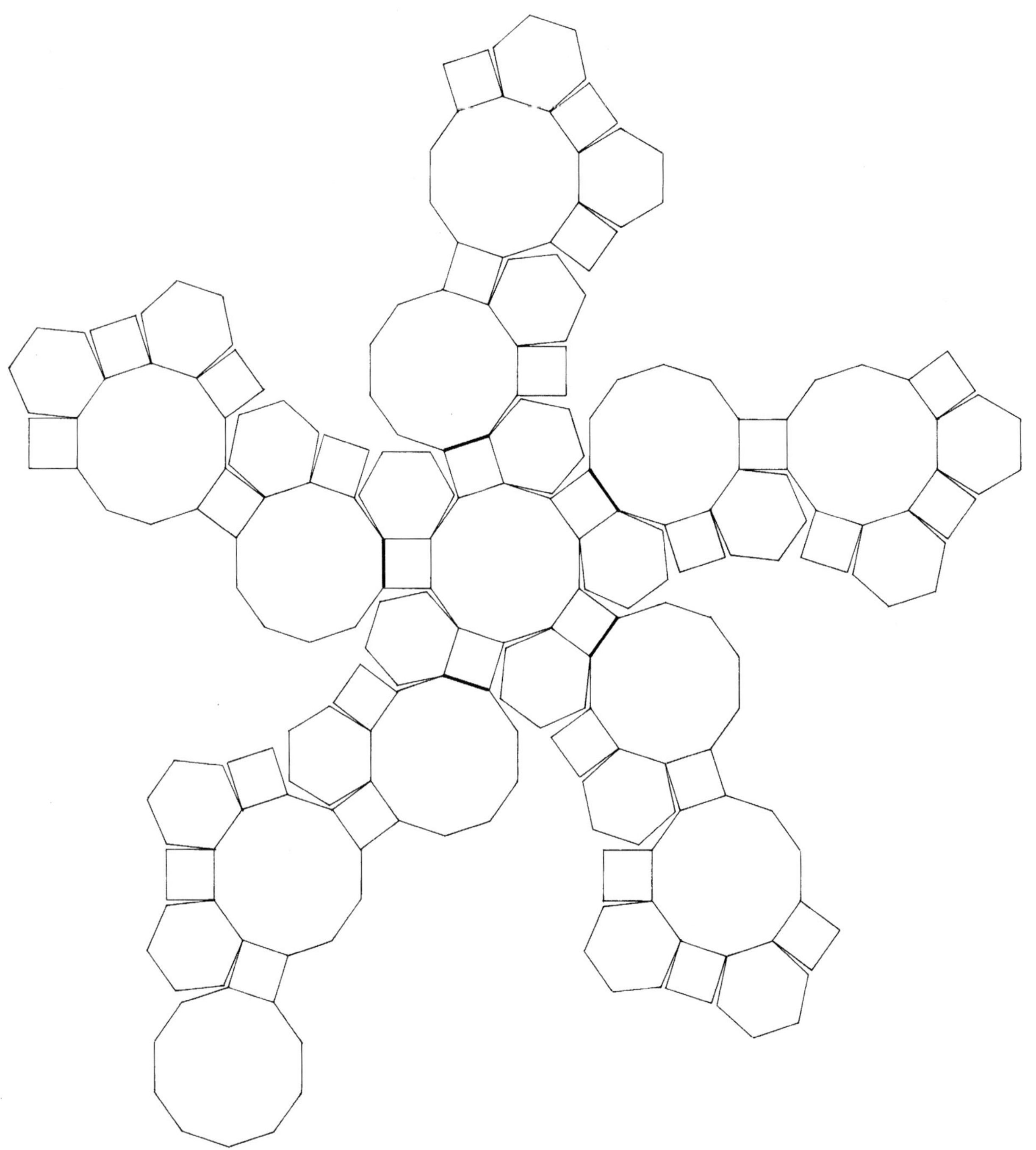

Fig. 8.17. Net of the Truncated Icosidodecahedron.

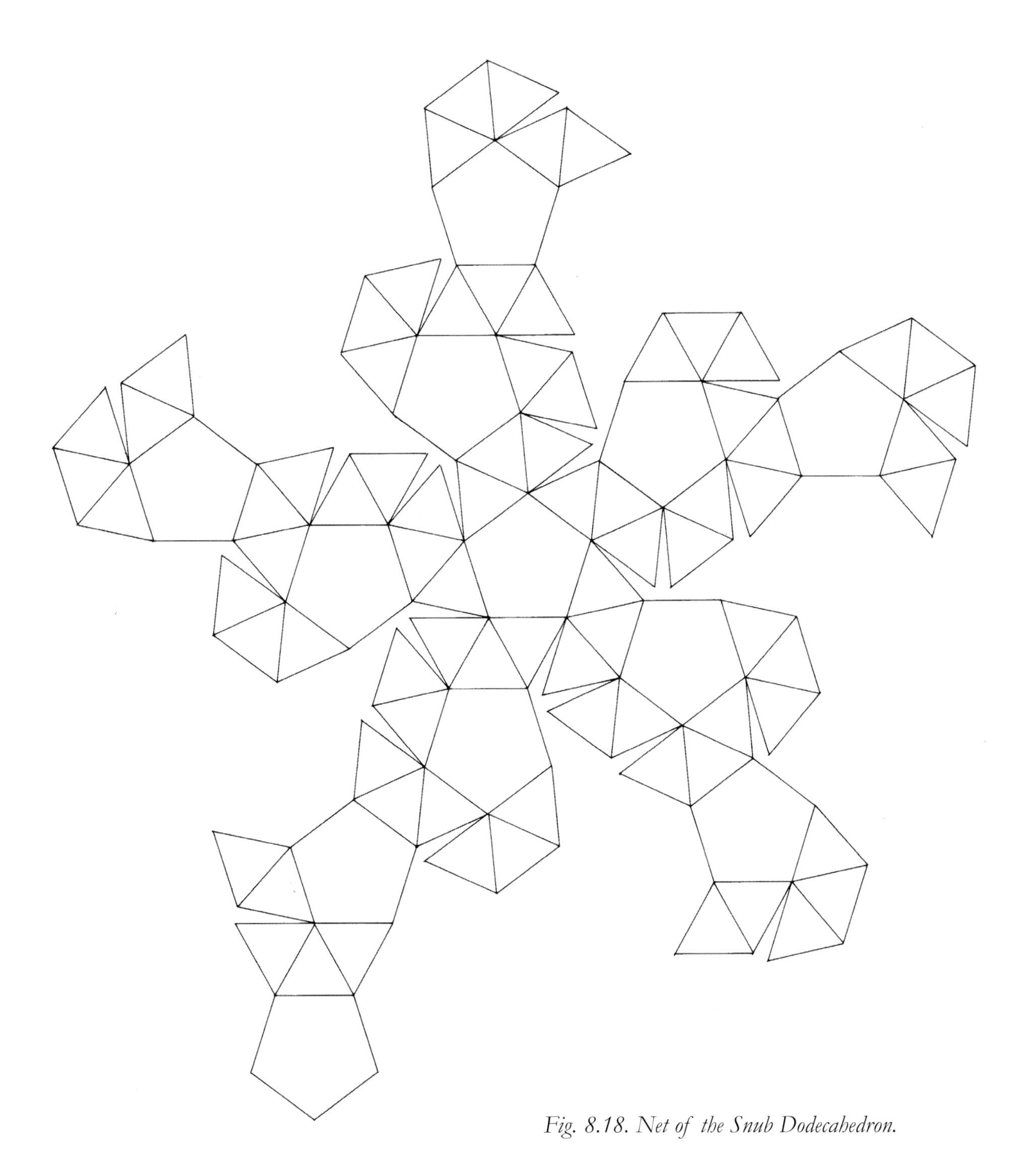

Fig. 8.18. Net of the Snub Dodecahedron.

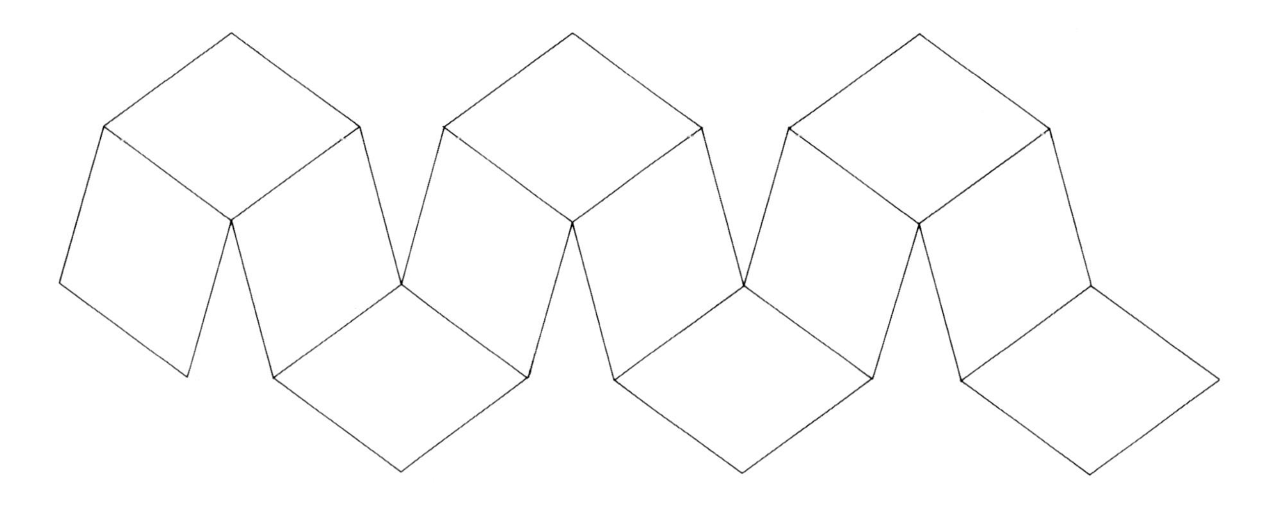

Fig. 8.19. Net of the Rhombic Dodecahedron.

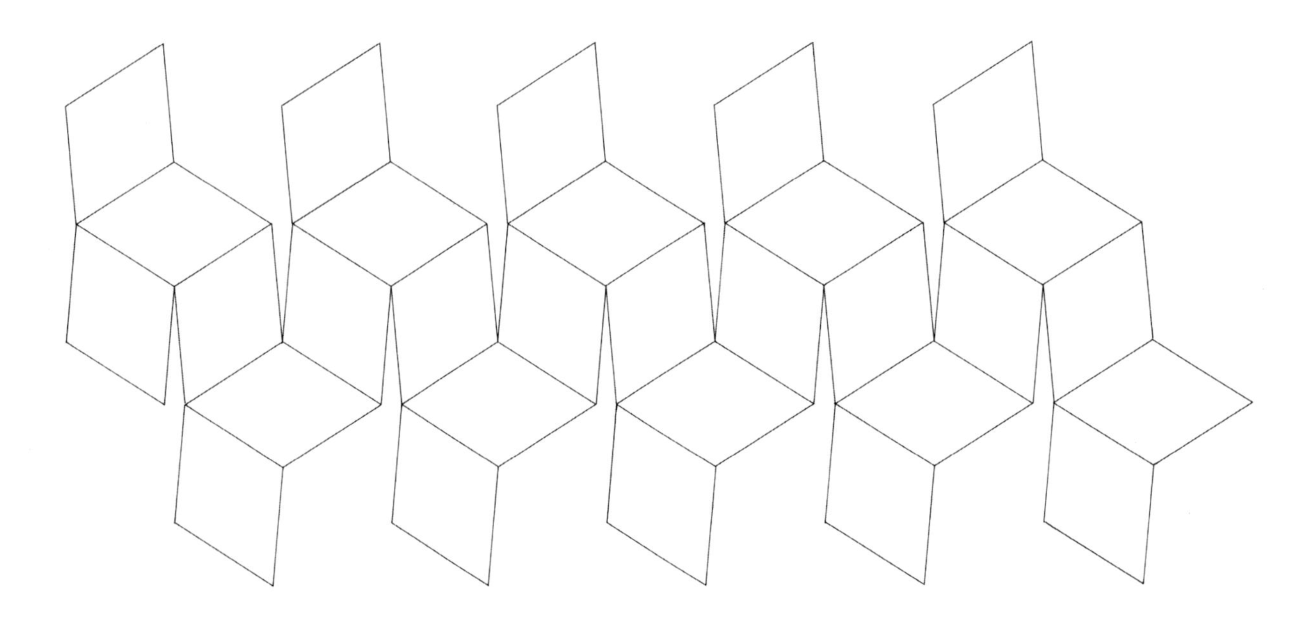

Fig. 8.20. Net of the Rhombic Triacontahedron.

Appendix C:
Data Table for the Platonic and Archimedean Solids

The Circumradius of a polyhedron is the radius of the sphere that exactly encompasses the figure (touching all vertices).

	Edge length (if Circumradius = 1)	Circumradius (if Edge length = 1)
Platonic Regular Solids		
Cube	1.15470	0.86602
Tetrahedron	1.63299	0.61237
Octahedron	1.41421	0.70710
Icosahedron	1.05146	0.95105
Dodecahedron	0.71364	1.40127
Archimedean Semi-Regular Solids		
Truncated Tetrahedron	0.85280	1.17260
Truncated Octahedron	0.63245	1.58115
Cuboctahedron	1.00000	1.00000
Truncated Cuboctahedron	0.43148	2.31760
Snub Cube	0.74420	1.34372
Rhombicuboctahedron	0.71481	1.39897
Truncated Cube	0.56217	1.77882
Truncated Icosahedron	0.40355	2.47800
Icosidodecahedron	0.61803	1.61804
Truncated Icosidodecahedron	0.26299	3.80242
Snub Dodecahedron	0.46386	2.15582
Rhombicosidodecahedron	0.44784	2.23294
Truncated Dodecahedron	0.33676	2.96947

Appendix D:
Recommended Reading and Resources

Companion volume

Jon Allen (2007) *Drawing Geometry: a primer of basic forms for artists, designers and architects* (foreword by Keith Critchlow), Floris Books, Edinburgh.

Classics

Critchlow, K.B. (1969) *Order in Space*, Thames & Hudson, London.

Lawlor, R. (1982) *Sacred Geometry*, Thames & Hudson, London.

Martineau, J. (2001) *A Little Book of Coincidence*, Wooden Books, Presteigne.

Other recommended reading

Cromwell, P.R. (1997) *Polyhedra*, Cambridge University Press, Cambridge.

Cundy, H.M., Rollett, A.P. (third edition 1981) *Mathematical Models*, Tarquin Publications, St Albans.

Sutton, D. (1998) *Platonic and Archimedean Solids*, Wooden Books, Presteigne.

Useful websites

www.georgehart.com
www.korthalsaltes.com
www.wholemovement.com

Supplies

You will get most of the materials you need from a good art or craft supplier (Atlantis in London is good for paper). The least easy may be the sticks for making the stick models, so here are a couple of useful sources:

John Bell & Croydon (150 x 2mm 'wooden applicators without swabs') www.johnbellcroydon.co.uk

Ashwood (150 x 2.5 mm bamboo skewers – just cut off the sharp end) www.ashwood.biz

Specialist Crafts Ltd (200 x 2.5 mm natural sticks and 200 x 2 mm coloured sticks) www.speccrafts.co.uk

Index

Air 32
Compounds 94
Critchlow, Keith 8
Cube 19, 24f, 36, 121
Cuboctahedron 48
Decagon 81
Dodecahedron 22, 25, 38, 121
Dymaxion 48
Earth 36
Ether 38
Euclid 100
Fire 30
Great Dodecahedron 105
Great Stellated Dodecahedron 108
Hexagon 65
Icosahedron 21, 25, 34, 120
Icosidodecahedron 71, 125
Octahedron 20, 24f, 32, 120
Pentagon 67
Plato 8
Pribilof Island 8
Pythagoras 8
Rhombic Dodecahedron 112, 132
Rhombic Triacontahedron 116, 132
Rhombicosidodecahedron 87, 129
Rhombicuboctahedron 58, 124
Small Stellated Dodecahedron 102
Snub 118
Snub Cube 87, 128
Snub Dodecahedron 90, 131
Square 49
Stellation 119
Symmetry 119
Tetrahedron 18, 24f, 30, 120
Triangle 49
Truncation 118
Truncated Cube 52, 123
Truncated Cuboctahedron 45, 124
Truncated Dodecahedron 74, 127
Truncated Icosahedron 67, 126
Truncated Icosidodecahedron 81, 130
Truncated Octahedron 45, 122
Truncated Tetrahedron 43, 122
Water 34
Wholeness 8

Contacting the Author

Jon Allen is a practising architect who uses geometry in his work to create beautiful and harmonious buildings.

He also teaches, and has contributed to a number of courses and conferences over the years. He is available to give workshops.

He is happy to receive feedback or questions regarding this book, and may be contacted via email: mail@jonallenarchitect.co.uk